Plumes
Delineation & Transport

by D. James Benton

Copyright © 2019 by D. James Benton, all rights reserved.

Preface

Contaminant plumes are most often considered, as these have some associated urgency. All of the plumes I have worked with in the past forty years have been either chemical or thermal and in some way impacted the environment. There are two main considerations for such plumes: 1) delineation and 2) transport. The associated questions are: 1) how much of what is present and 2) where is moving to. Combined with possible migration and spreading is the question of how it might be changing, including chemical and nuclear reactions. Whether the plume in consideration is in air, surface water, or ground water determines what calculations are appropriate, but the basic issues are the same in any case. In this text we will be considering all of these aspects with several actual plumes plus a variety of software, some of which is publicly available.

All of the examples contained in this book,
(as well as a lot of free programs) are available at…
https://www.dudleybenton.altervista.org/software/index.html

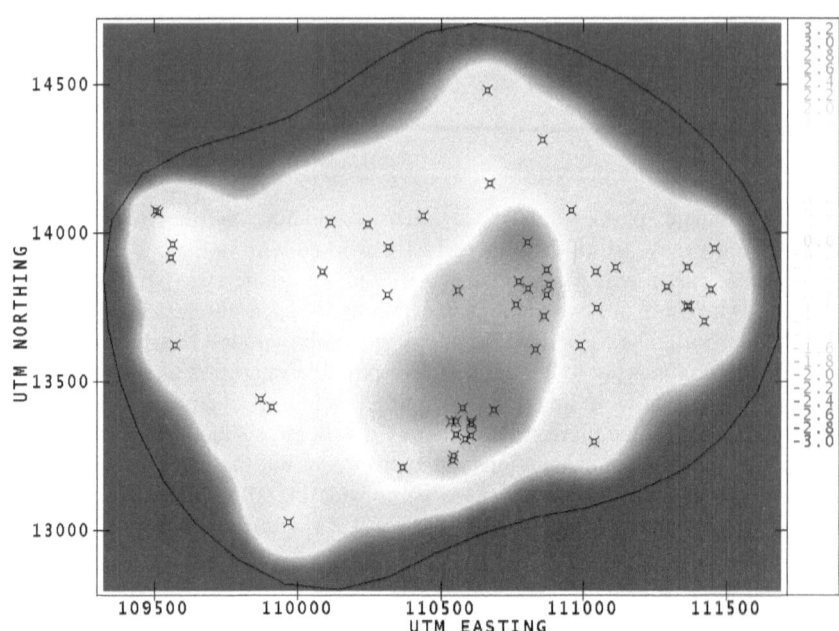

Typical 2D Contaminant Plume with Boundary (examples\plume1)

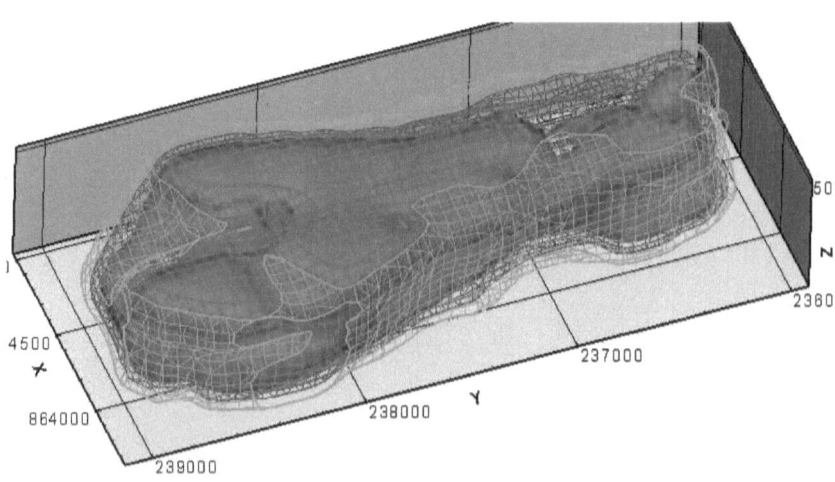

**Invariably, plumes get named. This one was named: *Drumstick*.
The bottom plume on the cover was named: *Sandal***

Table of Contents

	page
Preface	i
Chapter 1. Simple Airborne Contaminant	1
Chapter 2. Point Source Releases	3
Chapter 3. Dispersion	11
Chapter 4. Slot Jets	15
Chapter 5. 3D Thermal Plume	57
Chapter 6. Three-Dimensional Geological Data	65
Chapter 7. Contaminant Plumes in Groundwater	71
Chapter 8. Particle Tracking of Plumes	77
Appendix A. Inverse Distance Interpolation	93
Appendix B. Relaxation Method	97
Appendix C. Kriging	101
Appendix D. BMP to GIF Conversion	105

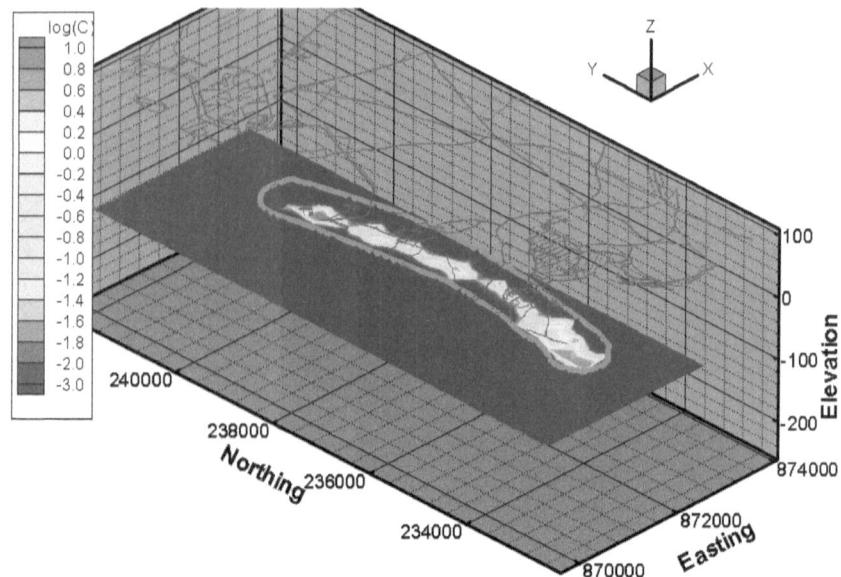

Horizontal slice through data along with magenta boundary.

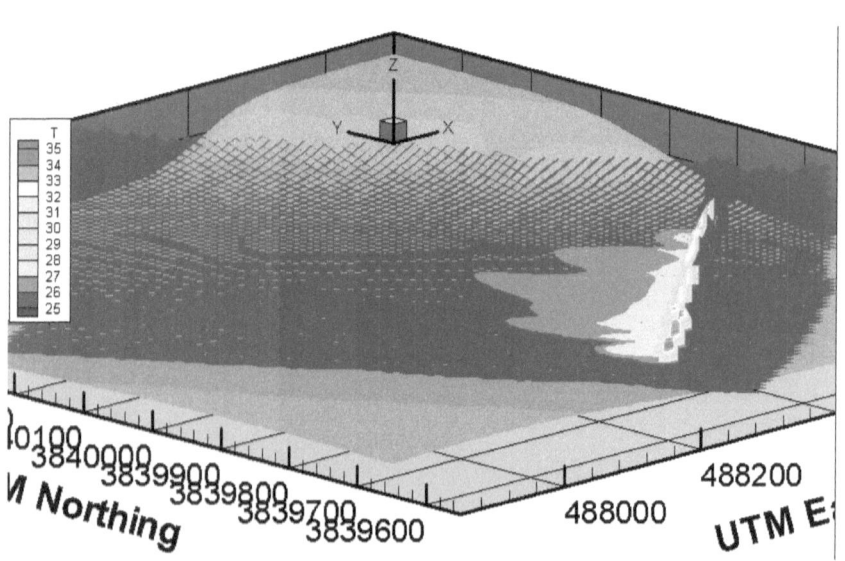

Thermal plume discharged into a river upward from the bottom.

Chapter 1. Simple Airborne Contaminant

The simplest plume I've ever been involved with modeling was ozone generated by some industrial processes, but mostly urban automotive exhaust around Birmingham, Alabama. The Bureau of Environmental Health had five sensors scattered around Jefferson and Shelby Counties, mostly along the corridor between I58 and I20. They provided some data and I developed software that would generate an animation of the estimated plume. A typical frame of this animation is shown below:

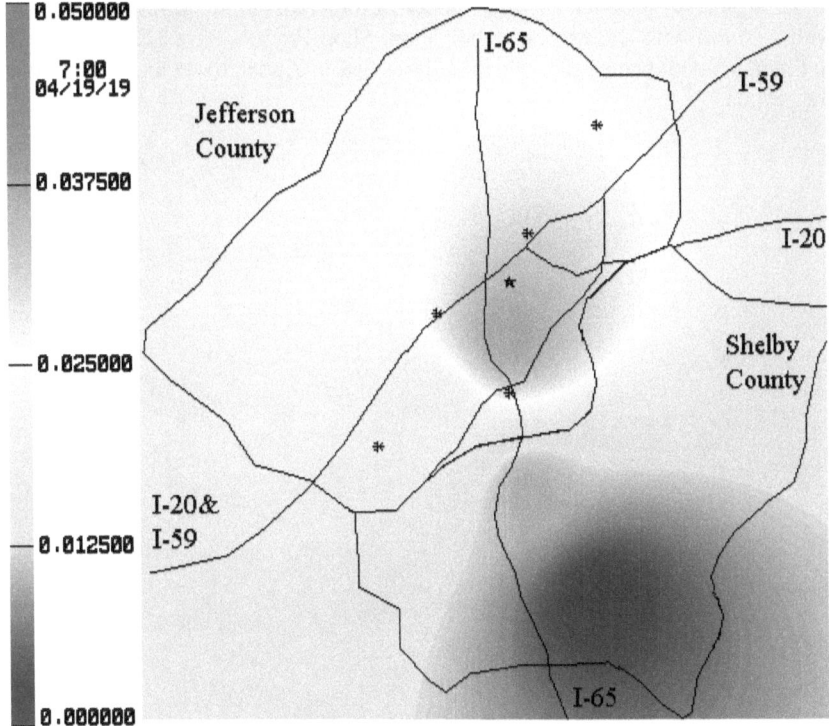

As this is very sparse data (only 5 sensors), there is no point applying elaborate processing. A simple inverse-distance weighting method for spatial interpolation of concentration is adequate. This can be expressed as:

$$C = \frac{\sum \dfrac{C_I}{R_I^{2.5}}}{\sum \dfrac{1}{R_I^{2.5}}} \quad (1.1)$$

A power of 2.5 on the distance is often applied. It is also necessary to control the concentrations in the far field (i.e., $\sim\infty$). This is easily done by adding a dozen phantom points at some distance outside the area of interest. See Appendix A for more on the inverse distance method. A more typical (and detailed) 2D contaminant plume is shown on page ii. Concentration in ppm is shown in the preceding figure, but log(C) is shown in the figure on page ii. Sometimes a log scale is more revealing, although when it comes to quantifying the contaminant, it is important to work with actual concentrations and not logs.

The code (ozone.c) produces a sequence of images (BMP files) representing the calculated spatial concentrations over time. These can be combined to form a single animation (GIF file) with a variety of tools, including Animation Shop®, which comes with the excellent tool, Paint Shop Pro®. I have also provided a rudimentary tool (bmp2gif), which is described in Appendix D and is available free online.

Chapter 2. Point Source Releases

The next level of detail we will consider are point source plumes. These are always approximate, as even a fissionable critical mass doesn't begin as an infinitesimal point. Point source plumes are the most common calculations for airborne contaminants. Interest in these calculations began in the early days of nuclear power when radioactive releases were feared imminent. When I worked at TVA I was part of a response team that would be called up in the event of a radioactive release. We had several such models ready, one for each nuclear plant and each release scenario.

We will begin by considering axisymmetric transient diffusion, as this is most often the assumption in these models. In one dimension this can be expressed by the following partial differential equation:

$$\frac{\partial C}{\partial t} = D\frac{\partial^2 C}{\partial x^2} \tag{2.1}$$

For our purposes here, the most logical boundary conditions are an initial contamination zone (a) and concentration (C_0) or total mass (m_0). The analytical solution in terms of the complementary error function is:

$$C(x,t) = \frac{C_0}{2}\left[erfc\left(\frac{x-a}{\sqrt{4Dt}}\right) - erfc\left(\frac{x+a}{\sqrt{4Dt}}\right)\right] \tag{2.2}$$

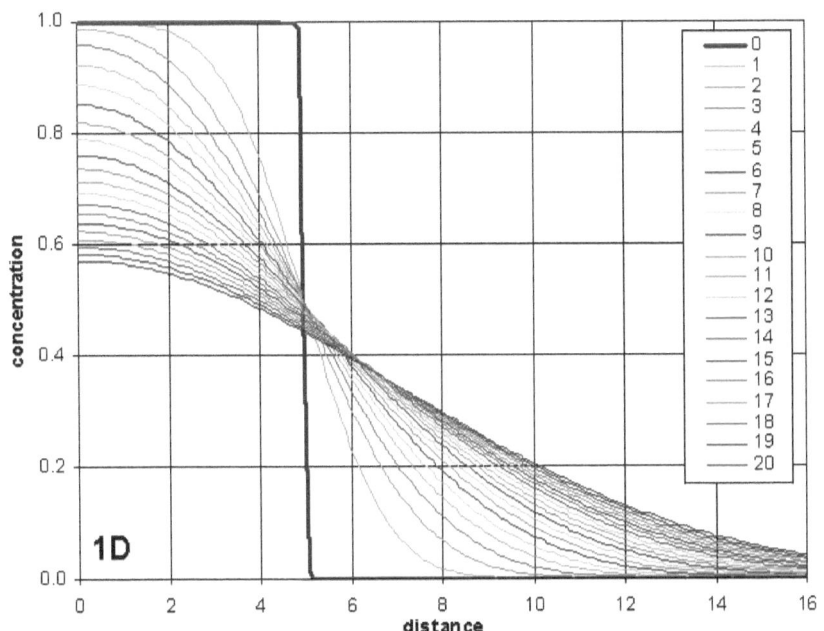

This solution is illustrated for several values of distance and time in the preceding figure. The area under each of the curves is the same so that the total mass is constant over time as it spreads in one direction.

$$m_0 = \int_0^\infty \frac{C_0}{2}\left[erfc\left(\frac{x-a}{\sqrt{4Dt}}\right) - erfc\left(\frac{x+a}{\sqrt{4Dt}}\right)\right]dx = C_0 a \qquad (2.3)$$

We can easily add a constant horizontal velocity (v) to this solution by simply substituting $x=x-vt/a$, as in:

$$C(x,t) = \frac{C_0}{2}\left[erfc\left(\frac{x-\frac{vt}{a}-a}{\sqrt{4Dt}}\right) - erfc\left(\frac{x-\frac{vt}{a}+a}{\sqrt{4Dt}}\right)\right] \qquad (2.4)$$

The resulting profiles move to the right, as illustrated below:

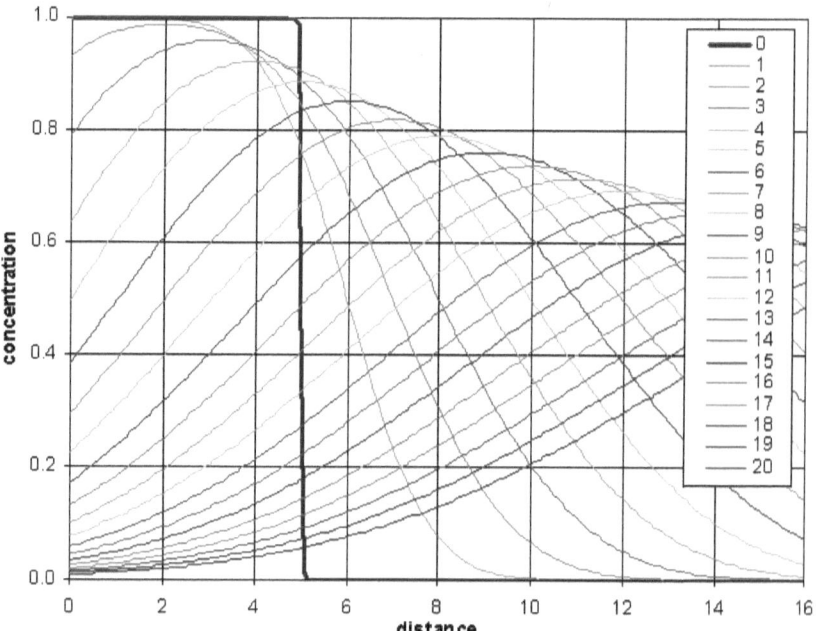

While this is interesting, no contaminant spreads in only one direction unless it's contained inside a pipe. We next consider two-dimensional cylindrical coordinates. Equation 2.1 becomes:

$$\frac{\partial C}{\partial t} = \frac{D}{r}\frac{\partial(rC)}{\partial r^2} \qquad (2.5)$$

The analytical solution to Equation 2.5 is an infinite series of Bessel functions, which is impractical for point source plume modeling, especially when such calculations were first undertaken in the late 1970s and early 1980s. We can obtain an approximate solution with realistic behavior by substituting $x = r^2/a$ into Equations 2.2 or 2.4. The resulting profiles are shown below:

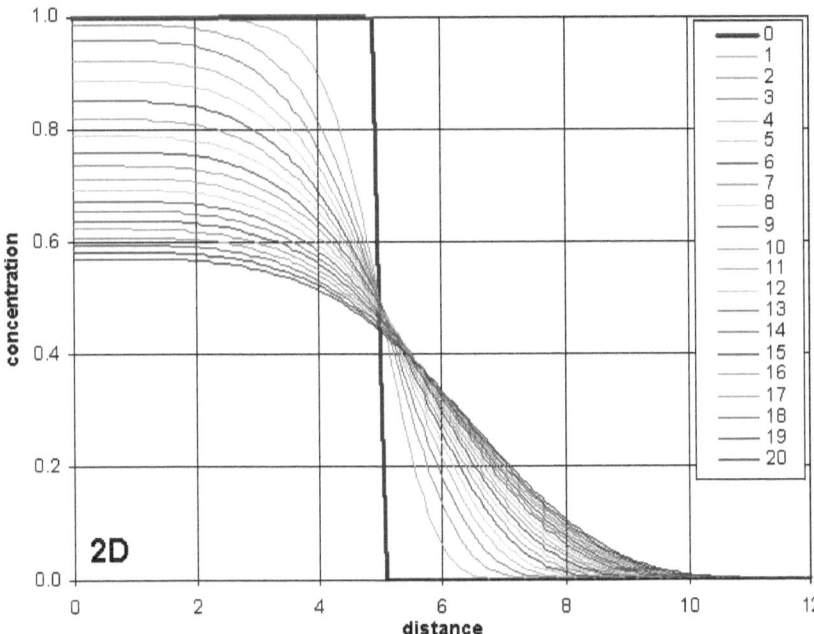

The conservation of mass in this case is given by:

$$m_0 = \int_0^\infty 2\pi C(t,r) r \, dr = \pi a^2 C_0 \qquad (2.6)$$

With this approximation, spreading increases with time, that is, the radius increases and the area under the curve remains constant. The maximum concentration, which is at the center in the absence of transverse velocity, also decreases with time, though much more slowly. The extent of the plume (or the leading edge where the concentration has dropped by three or four orders of magnitude) can be calculated using a bisection search. The VBA code is in the spreadsheet and also listed below:

```
Function radius(t As Double, c As Double, D As Double, a
    As Double) As Double
    Dim iter As Integer, c0 As Double, r1 As Double, r2 As
    Double
    c0 = conc(radius, t, c, D, 0, a)
    r1 = a
    r2 = a
```

```
While (conc(r2, t, c, D, 0, a) > c0 / 1000)
   r2 = r2 * 2
Wend
For iter = 1 To 32
   radius = (r1 + r2) / 2
   If (conc(radius, t, c, D, 0, a) > c0 / 1000) Then
      r1 = radius
   Else
      r2 = radius
   End If
Next iter
End Function
```

The results are shown in this next figure:

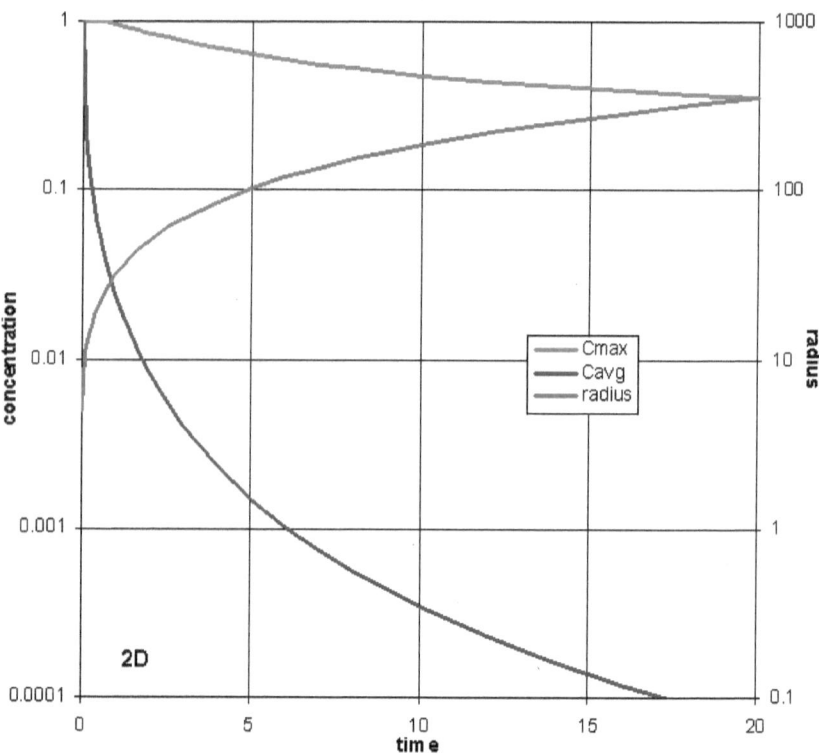

In three-dimensional spherical coordinates, this becomes:

$$\frac{\partial C}{\partial t} = \frac{D}{r^2}\frac{\partial (r^2 C)}{\partial r} \qquad (2.7)$$

The spherical solution is also an infinite series and impractical for our purposes. Fortunately, substituting $x=r^3/a^2$ into Equations 2.2 or 2.4 works again. The conservation of mass in 3D is:

$$m_0 = \int_0^\infty 4\pi C(t,r) r^2 \, dr = \frac{4\pi a^3}{3} C_0 \qquad (2.8)$$

The resulting profiles are shown in this next figure. All of the calculations can be found in the online archive in folder examples\point in spreadsheet point_source.xls, including the conservation of mass, which is along the bottom (i.e., row 163).

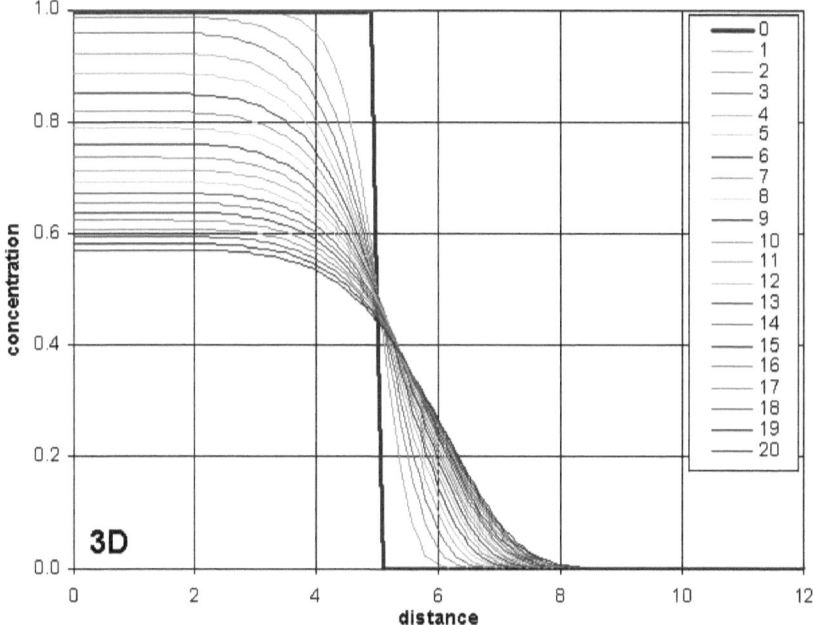

We can now use the modified Equation 2.4 to approximate an expanding point source 2D or 3D plume subject to a continuous cross wind, which is the first-order estimate of airborne contaminant transport.

Example 1

Consider an airborne plume having initial concentration of 1000 ppm and initial radius of 100 m. The diffusivity coefficient might be approximately 0.2 m²/s and the wind speed 0.8 m/s. This is a gentle breeze, not a howling wind. As a rough estimate, the plume spread might be mostly radial, that is, cylindrical instead of spherical. If this were the case, the average concentration would drop with the square of the radius. The peak concentration would also drop, but slower, as indicated in the preceding figure marked 2D.

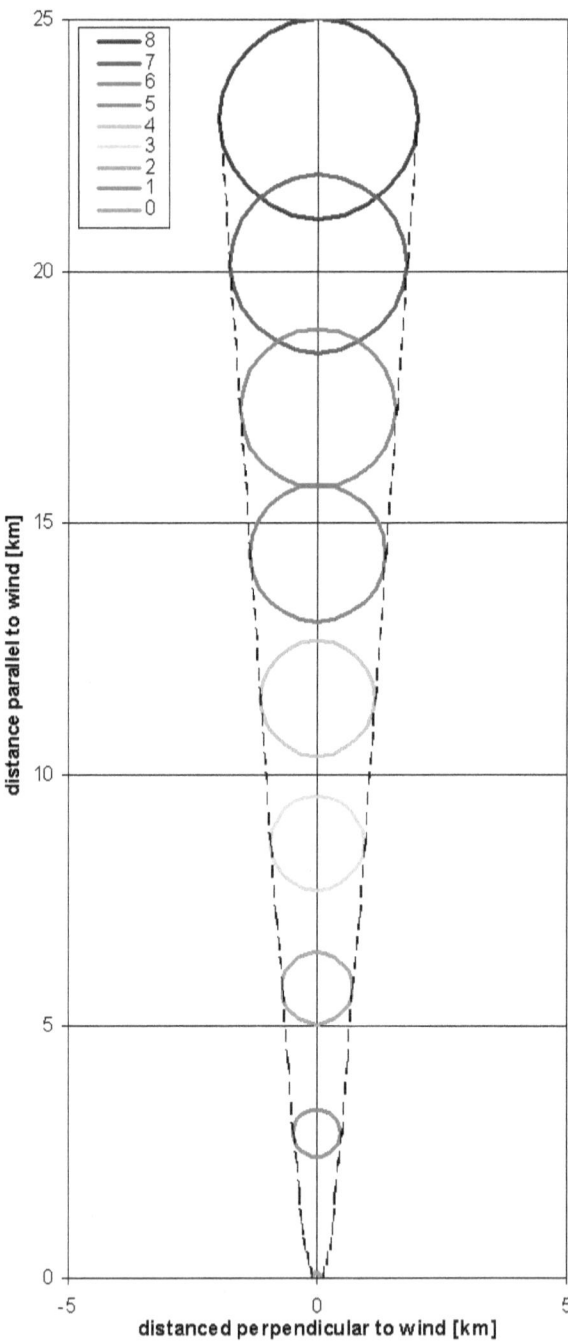

In this simplified calculation, the shift in center of the plume varies linearly with time (i.e., *x=vt*). The radial extent of the plume could be found in several ways, the easiest being a bisection search for the radius where the concentration has dropped of by a factor of 1000 or 0.1% of the maximum. It is easy enough to calculate the center and radius over time in a spreadsheet and produce a plot of the spreading, moving plume. Of course, this is an ideal representation and a more realistic approach would consider turbulent mixing, at least empirically.

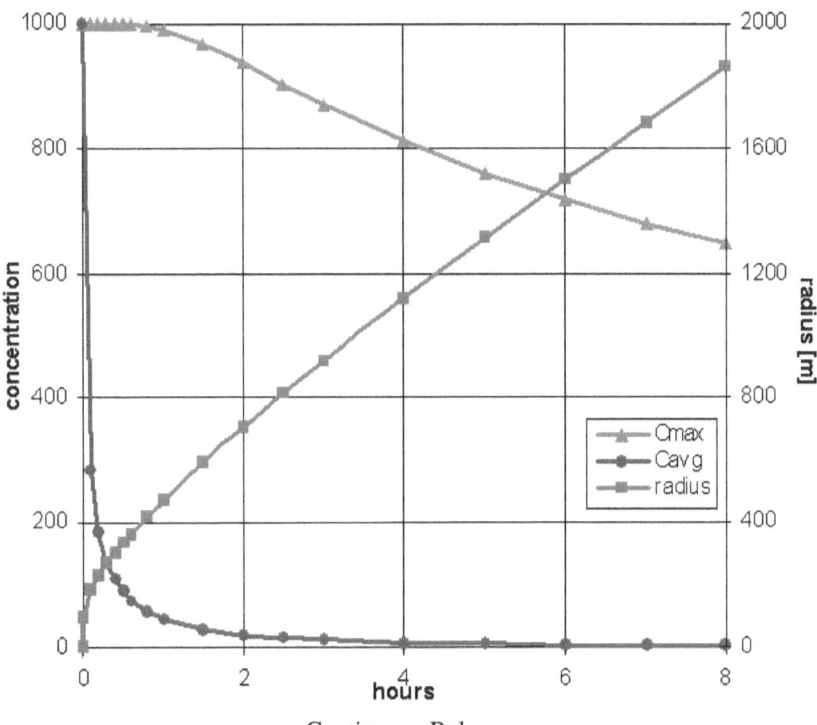

Continuous Releases

If you search the Internet for atmospheric plume modeling, you will find many articles that seem to be related to the topic of finite point source releases, but are not. The vast majority of those articles are about continuous (i.e., semi-infinite) releases. You will not find time as a variable in any of the equations presented therein. You may find a statement that those efforts are related to steady-state conditions. We will cover steady-state plumes later in the text. For now, we will only consider transient plumes.

Chapter 3. Dispersion

Diffusion and diffusivity generally describe a slow and continuous process, while dispersion and dispersivity often describe a more chaotic or turbulent process. The length scales for these two processes may be quite different. While diffusion arises from microscopic, even molecular processes, dispersion arises from macroscopic processes. Diffusion might be illustrated by the following:

which might be approximated by the familiar bell-shaped (Gaussian) curve:

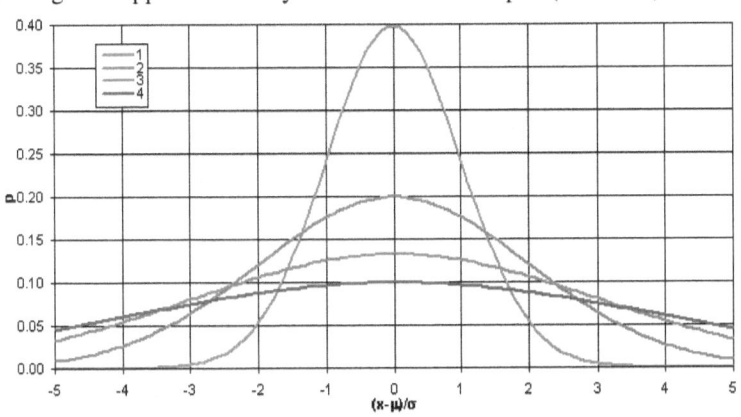

Dispersion is better illustrated by this figure, which defies analytical description:

The only way to generate concentrations anything like the preceding figure is with computer modeling: finite differences, finite elements, finite volume, or particle tracking. By far, the last of these is the most efficient and effective. We will not cover the specifics of the particle tracking model here, as this is thoroughly described in my book, *Particle Tracking*. Results for a groundwater contaminant plume much like the ones presented in the preceding chapter are shown in the following figure. The upper region is after 15 years and the lower one is after 30 years. Agreement with approximate analytical solutions was quite good, which was an important part of the validation process.[1,2]

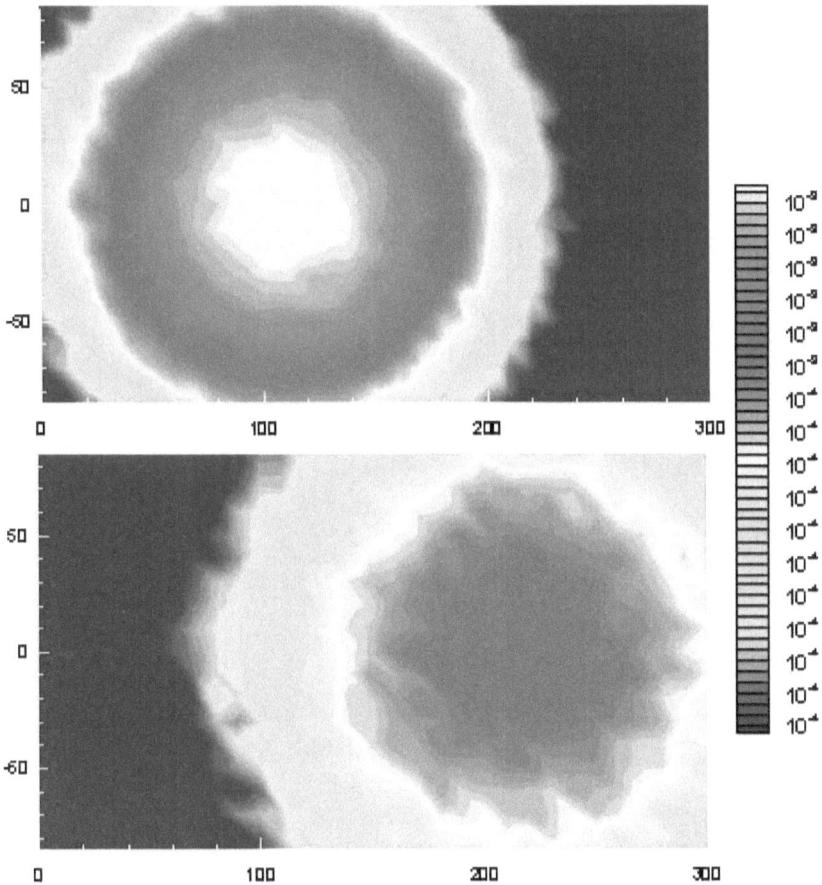

Analytical models just use a larger diffusion coefficient to approximate this behavior, but will always produce a smooth result unlike the one shown above. We will see much more realistic representations of dispersion in Chapter 8 with particle tracking.

---

[1] https://dudleybenton.altervista.org/pub/PTRAX1.pdf
[2] https://dudleybenton.altervista.org/pub/PTRAX2.pdf

Chapter 4. Slot Jets

The next problem we consider (a thermal plume discharged into a flowing river) is more complex because there are several equations to solve in parallel, including: conservation of mass, momentum, and energy, as well as width and position. There is also the conservation of salt if this is a brine plume or the receiving water is salty. The plume can be positively, negatively, or neutrally buoyant with respect to the ambient, and composed of heated or cooled fresh or salty water. The ambient can be stratified or uniform in temperature, flowing, stagnant, fresh or salty. The model is described in detail in my TWRA paper.[3]

At large power plants, thermal and wastewater effluents are frequently discharged into a receiving ambient through a diffuser. The primary purpose of this is to dilute the effluent. The most common diffuser shapes are conceptually similar to a solitary round jet or a linear cluster of jets, which ideally approaches a slot. The discharge is most often hotter or equal to the ambient in temperature. The discharge and the ambient can contain varying amounts of salt. The discharge can either rise, fall, spread, or some combination of these as it mixes with the ambient. The trajectory as well as the final dilution is of interest to the engineer. Analytical and empirical models are often adequate, even for discharge into a flowing ambient, but these can't effectively account for differences in temperature and salinity or variations in the ambient, including velocity, temperature, and salinity. In such cases a computer model is the best approach. A realistic computer model must be based on first principles, analysis, and observations.

Analytical And Empirical Models

Analytical and empirical models are briefly presented before the numerical model for three reasons: 1) the analytical and empirical models provide a check on the numerical model, especially the asymptotic behavior, 2) laboratory and field data are most often presented in the literature in conjunction with analytical or empirical models and this introduces the data for eventual comparison with the numerical model, and 3) the simplifying assumptions for the analytical and numerical models are basically the same. It will be shown that the analytical models have compared favorably with laboratory and field data, suggesting that the simplifying assumptions are not unreasonable. The numerical model handles slot and round jets; therefore, the analytical models for both geometries are included. Most laboratory and field data are either for a slot or a round jet; so including both geometries increases the data available for comparison.

Analytical models are limited to simple geometries and boundary conditions for which analytical solutions are possible. Empirical models are based on analytical models, dimensionless numbers, and data correlation. As with all

---

[3] Benton, D. J., "Development of a Two-Dimensional Plume Model for Positively and Negatively Buoyant Discharges into a Stratified Flowing Ambient," *Tennessee Water Resources Symposium*, 1989. https://dudleybenton.altervista.org/pub/Plume2D.pdf

models, some simplifications are necessary, in order to make the solution tractable, including: flow regime, geometry, and dominant mechanisms. Several factors are used to classify plumes, including: laminar or turbulent, round or slot, and buoyant or neutral. Simplifications are most often related to these classifications. The two limiting cases presented in analytical plume models are momentum-dominated and buoyancy-dominated.

Laminar Jet

Perhaps the most simple analytical model for a momentum-dominated plume was presented by Schlichting[4], clarified by Bickley[5], and described by White.[6] Schlichting derived an analytical solution for a neutrally-buoyant laminar slot jet discharging into a semi-infinite static medium. The dilution, S, is given by Equation 4.1:

$$S = 1 + \left(36Re\frac{h}{b}\right)^{1/3} \tag{4.1}$$

where h is the depth [m], b is the slot width [m], and Re is the slot Reynolds number, Equation 4.2:

$$Re = \frac{wb}{v} \tag{4.2}$$

where w is the velocity at the slot [m/s] and v is the kinematic viscosity [m²/s]. Note that the dilution depends on the Reynolds number and increases with the one-third power of depth. Even this simple analytical model has been shown to be reasonably accurate for Reynolds numbers up to 30, Andrade.[7] Although laminar flow and such small Reynolds numbers are not of direct interest, Schlichting's solution is important; because it provided the basis for solving a turbulent jet.

Squire[8] used Schlichting's laminar slot jet solution to develop a laminar round jet solution, Equation 4.3:

$$S = 1 + \frac{32}{Re}\left(\frac{h}{d}\right) \tag{4.3}$$

where d is the initial diameter of the jet [m] and Re is the jet Reynolds number, Equation 4.4:

---

[4] Schlichting, H., Zeitschrift für Angewandte Mathematik und Mechanik (Journal of Applied Mathematics and Mechanics), Vol. 13, pp. 260 263, 1933.
[5] Bickley, W. G., Philosophical Magazine, Vol. 23, pp. 727 731, 1937.
[6] White, F. M., Viscous Fluid Flow, McGraw-Hill, New York, pp. 290 292, 350 351, 1974.
[7] Andrade, E. N., Proceedings of the Physical Society of London, Vol. 51, pp. 784 793, 1939.
[8] Squire, H. B., Quarterly Journal of Mechanics, Vol. 4, pp. 321 329, 1951.

$$Re = \frac{wd}{v} \qquad (4.4)$$

Note that the dilution depends on the Reynolds number and increases linearly with depth.

Turbulent Jet

Prandtl[9] used his mixing length concept to modify Schlichting's laminar theory to account for turbulence. Prandtl's assumptions were later substantiated by Reichardt[10] and later Görtler[11] showed that the resulting equations could be solved by Schlichting's method, to derive Equation 4.5:

$$S = 1 + 0.48 \left(\frac{h}{b}\right)^{\frac{1}{2}} \qquad (4.5)$$

Note that the dilution does not depend on the Reynolds number and increases with the square-root of the depth. Görtler also developed a relationship for a round jet, Equation 4.6:

$$S = 1 + 0.424 \left(\frac{h}{d}\right) \qquad (4.6)$$

Note that the dilution does not depend on the Reynolds number and increases linearly with the depth.

Turbulent Plume

The analytical solution for a buoyancy-dominated plume discharged vertically upward from a slot at negligible velocity into a semi-infinite medium was developed by Rouse et al.[12], applied by Cederwall[13], and can be expressed by Equation 4.7:

---

[9] Prandtl, L, Proceedings of the Second International Congress on Applied Mechanics, Zurich, pp. 62 75, 1926.
[10] Reichardt, H., "Gesetzmässigkeiten der freien Turbulenz (Regularities of Free Turbulence)," Verein Deutscher Ingenieure (Association of German Engineers), Forschungs p. 414 (Research Paper No. 414), 1942.
[11] Görtler, H., "Berechnung von Aufgaben der freien Turbulenz auf Grund eines neuen Näherungsansatzes (The Task of Computating Free Turbulence Due to the Proximity of a Beginning)," Zeitschrift für Angewandte Mathematik und Mechanik (Journal of Applied Mathematics and Mechanics), Vol. 22, No. 5, pp. 244-254, 1942.
[12] Rouse, H., C. S. Yih, and H. W. Humphreys, "Gravitational Convection from a Boundary Source," Tellus, Vol. 14, 1952.
[13] Cederwall, K., "Gross Parameter Solutions of Jets and Plumes," ASCE Journal of the Hydraulics Division, Vol. 101, No. HY5, 1975.

$$S = 1 + \frac{0.59}{F_D^{\frac{2}{3}}}\left(\frac{h}{b}\right) \qquad (4.7)$$

where F_D is the slot densimetric Froude number, Equation 4.8:

$$F_D = \frac{w}{\sqrt{gb\left(\frac{\rho_A - \rho_P}{\rho_A}\right)}} \qquad (4.8)$$

where g is the acceleration of gravity [m/s²], ρ_A is the ambient density [kg/m³], and ρ_P is the initial plume density [kg/m³]. Note that the dilution depends on the Froude number, not the Reynolds number, and increases linearly with depth. Fischer et al.[14] develop a similar relation for a round plume, Equation 4.9:

$$S = 1 + \frac{0.117}{F_D^{\frac{2}{3}}}\left(\frac{h}{d}\right)^{\frac{5}{3}} \qquad (4.9)$$

where F_D is the jet densimetric Froude number, Equation 4.10:

$$F_D = \frac{w}{\sqrt{gd\left(\frac{\rho_A - \rho_P}{\rho_A}\right)}} \qquad (4.10)$$

Note that the dilution depends on the Froude number, not the Reynolds number, and increases with the five-thirds power of the depth.

Buoyant Jet

Actual discharges are typically somewhere between a momentum-dominated jet and a buoyancy-dominated plume. Fischer et al. provide a comparison between measured data and the analytical expressions for a round turbulent jet (Equation 4.6) and a round turbulent plume (Equation 4.9). This transformation of variables allows jet and plume data to be plotted on the same curve as shown in the following figure:

---

[14] Fischer, H. B., E. J. List, J. Imberger, and N. H. Brooks, Mixing in Inland and Coastal Waters, Academic Press, San Diego, California, 1979.

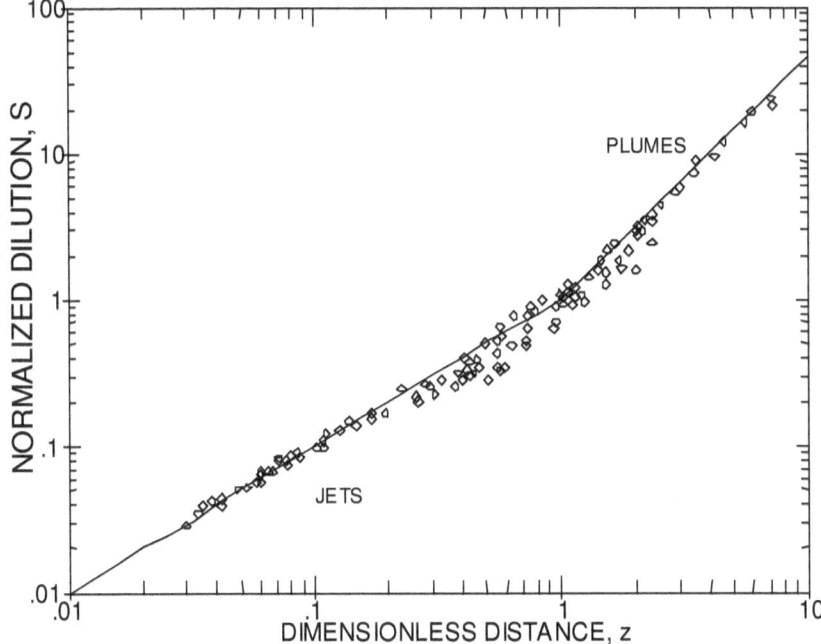

- The rearranged analytical model is given in Equation 4.11.

$$\varsigma = \begin{cases} \delta, & \delta \leq 1 \\ \delta^{\frac{5}{3}}, & \delta > 1 \end{cases} \qquad (4.11)$$

where δ is the dimensionless depth, Equation 4.12:

$$\delta = \frac{0.484}{F_D}\left(\frac{h}{d}\right) \qquad (4.12)$$

and ς is the normalized dilution, Equation 4.13:

$$\varsigma = \frac{1.69}{F_D}(S-1) \qquad (4.13)$$

<u>Inclined Jet</u>

The initial angle of inclination of a discharge has a pronounced impact on the trajectory of a plume and a lesser impact on the dilution. The impact of inclination angle has received much less attention in the literature than the other aspects presented here. One of the few publications containing laboratory data for a jet with variable inclination is that of Albertson et al.[15] and is reproduced in the next figure.

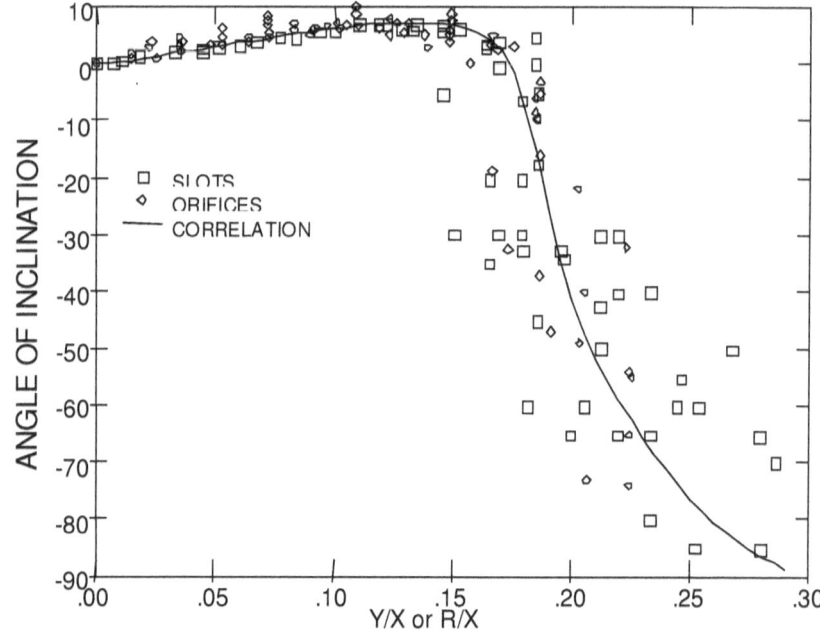

As expected, the greatest impact on the trajectory occurs when directing a buoyant discharge downward (-90°). The above curve can be calculated by Equation 4.14:

$$\alpha = \frac{-15.3*\beta \, (5.78*\beta-1) \, (231*\beta+1) \, (117*\beta^2-13.1*\beta+1) \, (32.2*\beta^2-11.1*\beta+1) \, (8.94*\beta^2-5.9*\beta+1)}{(28.5*\beta^2-10.6*\beta+1)}$$

(4.14)

where $\beta=y/x$ for slot jets and $\beta=r/x$ for round jets. Equation 4.14 is implicit in α and explicit in β as more than one angle of inclination can result in the same

---

[15] Albertson, M. L., Y. B. Dai, R. A. Jensen, and H. Rouse, "Diffusion of Submerged Jets," Transactions of the American Society of Civil Engineers, Paper No. 2409, 1948.

impact ratio. Even though Equation 4.14 is a high-order curve-fit it has been arranged so as to have only two real roots and stable behavior even with limited precision calculations.

Impact of Crossflow

Discharge into a flowing ambient is an important aspect of most jets and plumes. The simplest and most analyzed type of ambient is a uniform crossflow. Purely analytical solutions to this problem have remained illusive; but empirical solutions have been developed. Discharges may behave like a jet in the near field and like a plume in the far field, reacting differently to a crossflow throughout its trajectory. In some cases, for example Wright[16], List[17], and Wood et al.[18], four or more separate empirical relationship are developed, one for each regime: near field momentum-dominated, near field buoyancy-dominated, far field momentum-dominated, far field buoyancy-dominated, etc.

Perhaps the simplest general approach for a momentum-dominated *slot* discharge into a uniform crossflow is given by Adams[19], which can be expressed as Equation 4.15:

$$S = \frac{1}{2}\left(\frac{Q_R}{Q_D} + \left[\left(\frac{Q_R}{Q_D}\right)^2 + 2\frac{h}{b}\right]^{\frac{1}{2}}\right) \quad (4.15)$$

where h is the ambient depth [m], Q_D is the discharge flow [m³/s], and Q_R is the ambient (or river) flow [m³/s] over the diffuser. A comparison between Equation 4.15 and data can be seen in this next figure, which shows laboratory data from McIntosh et al.[20] and Seo et al.[21] as well as field data from Harleman et al.[22], Almquist et al.[23], and McIntosh et al.[24].

---

[16] Wright, S., "Mean Behavior of Buoyant Jets in a Crossflow," ASCE Journal of Hydraulics, Vol. 103, pp. 499-513, 1977.

[17] List, E. J., "Turbulent Jets and Plumes," Annual Review of Fluid Mechanics, Vol. 14, pp. 189-212, 1982.

[18] Wood, I., R. Bell, and D. Wilkinson, "Ocean Disposal of Wastewater," Advanced Series on Ocean Engineering, Vol. 8, World Scientific Publishers, Singapore, 1993.

[19] Adams, E. E., "Submerged Multiport Diffusers in Shallow Water with Current," Master's Thesis, Ralph M. Parsons Laboratory for Water Resources and Hydrodynamics, Department of Civil Engineering, MIT, Cambridge, Massachusetts, 1972.

[20] McIntosh, D. A., B. E. Johnson, and E. B. Speaks, "Validation of Computerized Thermal Compliance and Plume Development at Sequoyah Nuclear Plant," TVA Report WR28 1 45 115, 1983.

[21] Seo, I. W., D. S. Kim, and H. S. Kim, "The Mechanics of Tee Diffuser for Thermal Discharges in Crossflow," 4th International Conference on Hydro-Science and - Engineering, 2000.

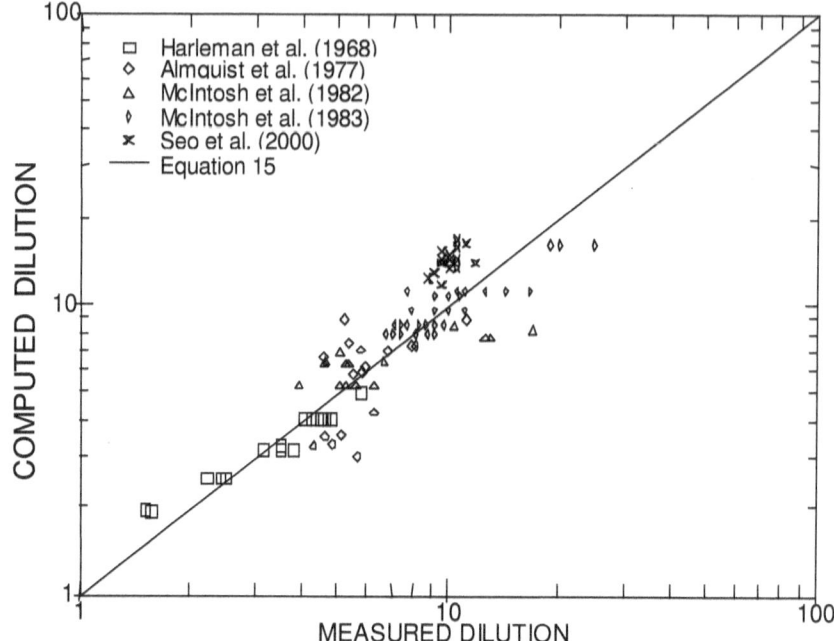

Harleman et al. suggest a simple formula for the dilution of buoyant *slot* plumes in a crossflow:

$$Q_F = 0.56 \left[g D^3 L^2 Q_D \left(\frac{\rho_A - \rho_D}{\rho_A} \right) \right]^{\frac{1}{3}} \quad (4.16a)$$

$$Q_M = \frac{1}{2} \left(Q_R + \sqrt{Q_R^2 + \frac{5D\,Q_D}{3L}} \right) \quad (4.16b)$$

---

[22] Harleman, D. R. F., L. C. Hall, and T. G. Curtis, "Thermal Diffusion of Condenser Water in a River During Steady and Unsteady Flows," Hydrodynamics Laboratory Report No. 111, Massachusetts Institute of Technology, Cambridge, Massachusetts, 1968.

[23] Almquist, C. W., C. D. Ungate, and W. R. Waldrop, "Field Model Results for Multiport Diffuser Plume," Proceedings of the ASCE Conference on Verification of Mathematical and Physical Models in Hydraulic Engineering, College Park, Maryland, 1978.

[24] McIntosh, D. A., B. E. Johnson, and E. B. Speaks, "A Field Verification of Sequoyah Nuclear Plant Diffuser Performance Model: One-Unit Operation," TVA Report WR28 1 45 110, 1982.

$$\text{if } (Q_R \geq Q_F) S = \frac{Q_F}{Q_D} \qquad (4.16c)$$

$$\text{if } (Q_R < Q_F) S = S = \frac{Q_M}{Q_D} \qquad (4.16d)$$

Equation 4.16 is introduced here because it will subsequently be compared to laboratory and field data. Stolzenbach[25] suggests a slightly more complex formula for the dilution of buoyant *slot* plumes in a crossflow:

$$Q_H = \frac{1}{2}\left(Q_R + \sqrt{Q_R^2 + \frac{10L}{D}Q_D^2}\right) \qquad (4.17a)$$

$$\text{if } (Q_R \geq Q_H) S = \frac{Q_H}{Q_D} \qquad (4.17b)$$

$$Q_F = D\left[L^2 Q_D (T_D - T_A)\frac{d\rho}{dT}\right]^{\frac{1}{3}} \qquad (4.17c)$$

$$\text{if } (Q_R \geq Q_F) S = \frac{Q_R}{Q_D} \qquad (4.17d)$$

$$\text{if } (Q_R < Q_F) S = \frac{Q_F}{Q_D} \qquad (4.17d)$$

Equation 4.17 is introduced here because it will subsequently be compared to laboratory and field data. Almquist suggests a set of formulae for the dilution of buoyant *slot* plumes in a crossflow that considers twelve different regimes based on the relative importance of buoyancy vs. momentum and positive, negative, and zero ambient flow:

$$S_{FBDZ} = 1 + \frac{C_B b_0^{\frac{1}{3}} h}{Q_Z} \qquad (4.18a)$$

$$\sigma = 1 + C_M \left(\frac{vh}{Q_Z} + \sqrt{\frac{v_2 h^2}{Q_Z^2} + \frac{2M_0 h}{Q_Z^2}}\right) \qquad (4.18b)$$

---

[25] Stolzenbach, K. D., "Analytical and Experimental Studies of Discharge Designs for the Cayuga Station at the Somerset Alternate Site," Ralph M. Parsons Laboratory for Water Resources and Hydrodynamics Report No. 211, 1976.

$$Q_B = (\sigma - 1)Q_Z L \qquad (4.18c)$$

$$Q_X = \frac{Q_R h_L}{h_1} \qquad (4.18d)$$

$$S_{FMDP} = 1 + \frac{Q_X}{Q_Z L} \qquad (4.18e)$$

$$S_{FMDE} = 1 + C_M \sqrt{\frac{2M_0 h}{Q_Z}} \qquad (4.18f)$$

$$S_{TRNZ} = 1 + \frac{C_T b_0^{\frac{1}{6}} M_0^{\frac{1}{4}} h^{\frac{3}{4}}}{Q_Z} \qquad (4.18g)$$

$$S_{TRNP} = S_{TRNZ} + (S_{FMDP} - S_{TRNZ}) v_B \qquad (4.18g)$$

$$Q_Z = \frac{Q_D}{L} \qquad (4.18h)$$

$$M_O = \frac{Q_Z^2}{0.3709} \qquad (4.18i)$$

$$b_0 = \frac{g(\rho_A - \rho_D)}{\rho_A} Q_Z \qquad (4.18j)$$

$$h_L = \frac{D b_0^{\frac{2}{3}}}{M_0} \qquad (4.18k)$$

$$v = \frac{Q_R}{LD} \qquad (4.18l)$$

$$v_B = \frac{3v}{2} b_0^{\frac{2}{3}} \qquad (4.18m)$$

$$m = \frac{v^2 h}{M_0} \qquad (4.18n)$$

$$b = g \rho Q_Z \qquad (4.18p)$$

where subscript B indicates buoyancy dominated, h is the depth, h_L is the effective mixing length, subscript M indicates momentum dominated plus dimensionless momentum ratio, L is the diffuser length, Q indicates discharge or river flow, depending on the subscript, subscript R indicates reverse flow, S is dilution the, subscript T indicates transition, and subscript Z indicates zero flow.

Equation 4.18 is introduced here because it will subsequently be compared to laboratory and field data and this set of formulae have special historical significance in the development of the plume model. Huang et al.[26] developed an expression for a *round* buoyant plume discharging into a uniform crossflow, which can be rearranged to form Equation 4.19:

$$S = \frac{4}{\pi}\left(\frac{h}{d}\right)^2\left(\frac{u}{w}\right)\left[0.1\zeta^{\frac{1}{3}} + \frac{0.51}{\left(1+0.1\zeta^2\right)}\right] \quad (4.19)$$

where ζ is the dimensionless depth, Equation 4.20:

$$\zeta = \left(\frac{\pi}{4F_D^2}\right)\left(\frac{d}{h}\right)\left(\frac{w}{u}\right)^3 \quad (4.20)$$

Agreement between Equation 4.19 and the data of Lee and Cheung[27] is shown in this next figure.

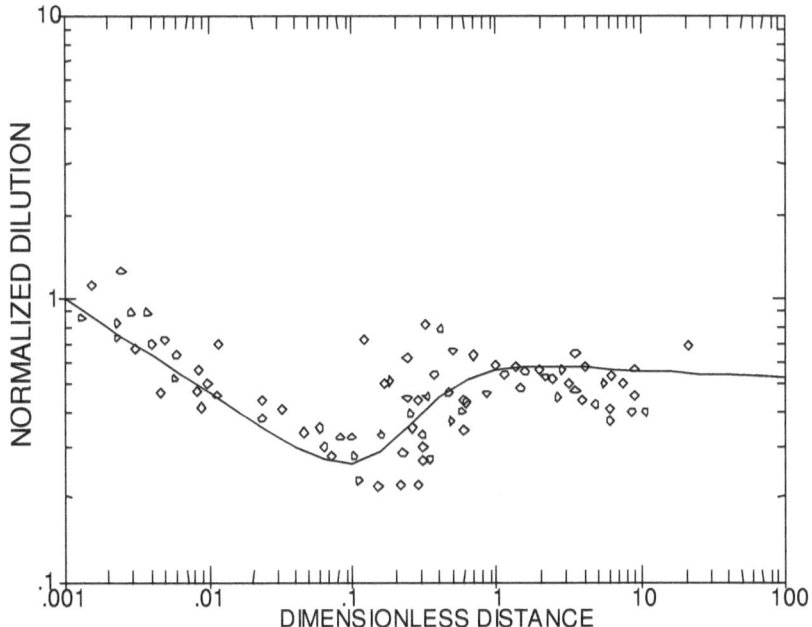

---

[26] Huang, H., R. Fergen, J. Proni, and J. Tsai, "Initial Dilution Equations for Buoyancy-Dominated Jets in Current," Journal of Hydraulic Engineering, Vol. 124, No. 1, pp. 105 108, 1998.

[27] Lee, J. H. W. and V. Cheung, "Generalized Lagrangian Model for Buoyant Jets in Current, ASCE Journal of Environmental Engineering, Vol. 116, No. 6, pp. 1085 1105, 1991.

Impact of Stratification

Stratification is present in the ambient, most often due to a temperature gradient. Several authors already cited have also dealt with this complication, for instance, Fischer, Wood, and their associates. Most of the papers on discharge into a stratified ambient have focused on buoyant plumes rising in an ambient with linearly varying density. The data and correlations are most often presented in the form of maximum height reached before stagnation. A straightforward approach

$$\frac{z}{d} = \frac{1}{F_D}\left[\frac{1.5}{J^{\frac{1}{4}}} + \frac{3.7}{J^{\frac{3}{8}}}\right] \quad (4.21)$$

yielding dilution is given by Wood et al.. Their correlations for maximum height reached by round and slot jets are given by Equations 4.21 and 4.22, respectively.

$$\frac{z}{b} = \frac{1}{F_{\mathcal{D}}^{\frac{4}{1}}}\left[\frac{4.6}{J^{\frac{1}{3}}} + \frac{0.67}{J^{\frac{1}{2}}}\right] \quad (4.22)$$

where J is the normalized stratification (similar to the densimetric Froude number) and is given by Equation 4.23:

$$J = -\frac{w^2 \frac{d\rho_A}{dz}}{g(\rho_A - \rho_P)} \quad (4.23)$$

The correlations of Wood et al. for dilution of round and slot jets are given by Equations 4.24 and 4.25, respectively.

$$S = \frac{1}{F_{\mathcal{D}}^{\frac{4}{1}}}\left[\frac{0.55}{J^{\frac{1}{4}}} + \frac{0.46}{J^{\frac{5}{8}}}\right] \quad (4.24)$$

$$S = \frac{1}{F_{\mathcal{D}}^{\frac{2}{1}}}\left[\frac{0.26}{J^{\frac{1}{6}}} + \frac{0.76}{J^{\frac{1}{2}}}\right] \quad (4.25)$$

Agreement between Equations 4.21 and 4.22 and the experimental data for maximum height reached of Wong[28] and Abraham and Eysink[29] are shown in this next.

---

[28] Wong, D.,, "Buoyant Jet Entrainment in Stratified Flows," Ph.D. Thesis, University of Michigan, Ann Arbor, Michigan 1986.

[29] Abraham, G., and W. Eysink, "Jets Issuing into a Fluid with a Density Gradient," Journal of Hydraulic Research, Vol. 7, pp. 145 147, 1968.

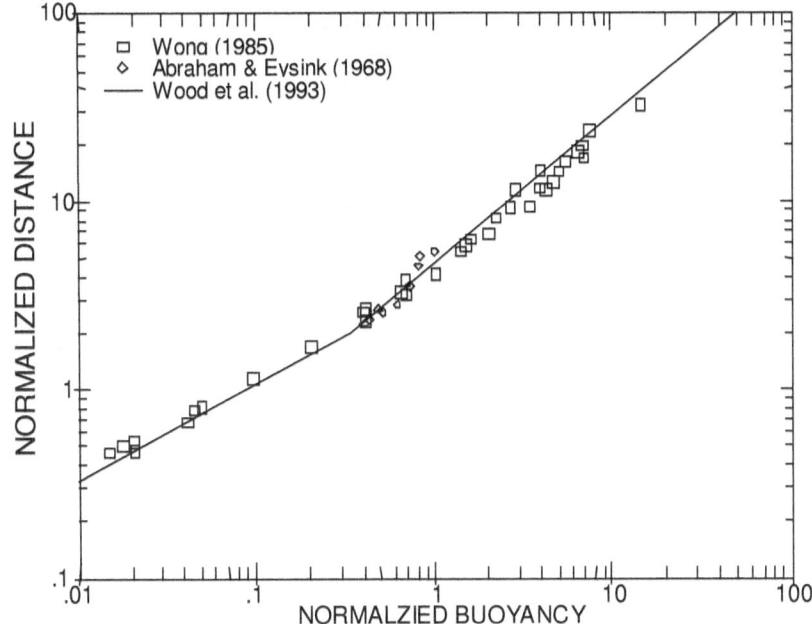

Agreement between Equations 4.24 and 4.25 and the experimental data for dilution of Wong are shown in this next figure.

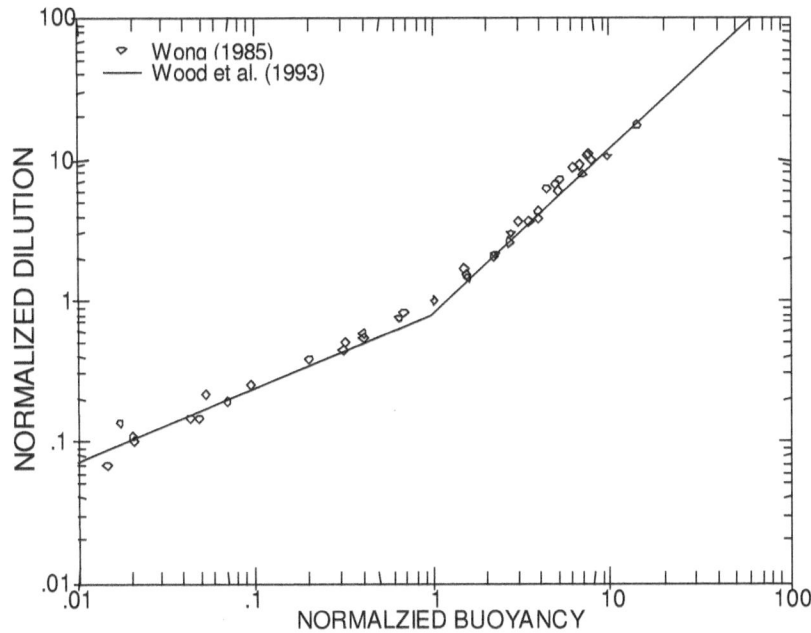

Mean Entrainment Temperature

Perhaps the simplest way to account for a thermally stratified ambient is to use some mean entrainment temperature. The rate of entrainment varies over the extent of the plume such that using a linear average of the ambient temperature does not yield sufficiently accurate results. The appropriate weighting with which to average the ambient temperature can be deduced from the analytical relationships already presented. Prandtl's relationship for a momentum-dominated or turbulent slot discharge, Equation 4.5, shows the dilution (and, thus, entrainment) increasing with the square root of the depth. Cederwall's relationship for a buoyancy-dominated or turbulent slot plume discharge, Equation 4.7, shows the entrainment increasing linearly with the depth. The relationship of Fischer et al., as shown in the first figure in this chapter, is consistent with these two, showing the entrainment for buoyancy-dominated plumes varying more steeply with depth than momentum-dominated jets. McIntosh suggests that for a slot discharge in general the dilution and entrainment increase with the three-fifths power of the depth. The weighting which, upon integration, results in the positive three-fifths power of depth is the negative two-fifths power, or Equation 4.26:

$$T_e = \frac{3}{5}\int_0^1 T\left(\frac{z}{h}\right)^{-\frac{2}{5}} d\left(\frac{z}{h}\right) \tag{4.26}$$

As this integral has a singularity at $z=0$, simplistic methods of numerical integration will not work; however, there are ways to work around this problem. A trapezoidal rule of integration using the average of each pair of $z's$ is adequate. Gauss quadrature is a common and accurate method of numerical integration which does not require the end point at $z=0$. If the temperature varies in some simple algebraic manner, such as linearly, an analytical solution can be obtained. A linearly varying temperature can be expressed by Equation 4.27:

$$T = T_o + (T_h - T_o)\left(\frac{z}{h}\right) \tag{4.27}$$

Substituting Equation 4.27 into Equation 4.26 and solving yields Equation 4.28:

$$T_e = T_o + \frac{3}{8}(T_h - T_o) \tag{4.28}$$

A simple arithmetic average would yield one-half instead of three-eights in Equation 4.28. For a linearly varying temperature this is equivalent to using the temperature at three-fifths of the total depth.

Jet Interference

Although the preceding analytical and empirical relationships suggest that the entrainment is greater for a round discharge than a slot discharge there are practicalities of scale and fabrication, which may not be reflected in the equations and must be considered. It is often impractical to design, construct, or operate a single round jet for a large discharge. One of the implicit assumptions in these analytical and empirical relationships is that the diffuser is small compared to the ambient and the laboratory models on which they are based are so constructed. If a prototype diffuser violates this assumption the actual performance may be significantly less than suggested by the scale model.

Multi-port diffusers are much more common for these reasons. A multi-port diffuser is easier to construct than a slot; but it is not a true slot nor a solitary jet, the geometric bases for developing the analytical and empirical relationships. Considerable attention has been given to the actual behavior of jets in close proximity, such as would be the case in a multi-port diffuser, which is typically a pipe with rows of round holes. Holley and Jirka[30] include a section entitled "Jet Interference" in which they discuss various aspects of the phenomenon and ways of accounting for it in a model. They basically suggest a distance at which the jets can be considered merged, which is about twice the distance between them, and an equivalent slot width, which may be based on a simple equivalent area or up to twice this width to account for diminished entrainment due to interference. Several three-dimensional computational fluid dynamics computer models of this jet interference and merging zone have been presented in the literature but these are beyond the scope of the present discussion.

Analytical and Empirical Model Assumptions

The following assumptions are implicit in developing all of these analytical and empirical models:

1) The discharge plume/jet is a coherent object, distinct from the ambient, which can be characterized by a profile or distribution and mean parameters.
2) The discharge plume/jet is small relative to the ambient such that it entrains the ambient, a process which changes its character but does not change the character of the ambient so much that the profile or distribution and mean parameters characterizing the ambient are no longer meaningful.
3) Interaction between the plume/jet and the ambient occurs at the boundary between the two and is limited to entrainment.
4) Small-scale turbulent phenomena can be characterized by bulk parameters such as mixing length and turbulent viscosity.
5) Large-scale turbulent phenomena can be averaged over time and accounted for by empirical factors such as entrainment coefficients.

---

[30] Holley, E. R., and G. H. Jirka, "Mixing in Rivers," Environmental and Water Quality Operational Studies Technical Report E 86 11, U. S. Army Corps of Engineers Waterways Experiment Station, Vicksburg, Mississippi, 1986.

Considering these assumptions and the associated simplifications involved in developing these models, the agreement shown in Figures 1 through 6 is remarkable. Besides providing a basis for testing the asymptotic behavior of the numerical plume model, these analytical solutions are offered as evidence that the assumptions yield acceptable results.

Numerical Model

The present numerical model was developed to address a particular need: dilution in a thermally stratified ambient, and subsequently expanded for other uses, for instance, discharge of brine. The formulation of the numerical model was influenced by this historical context. The numerical model is based on a concept, which can be illustrated by the following figure.

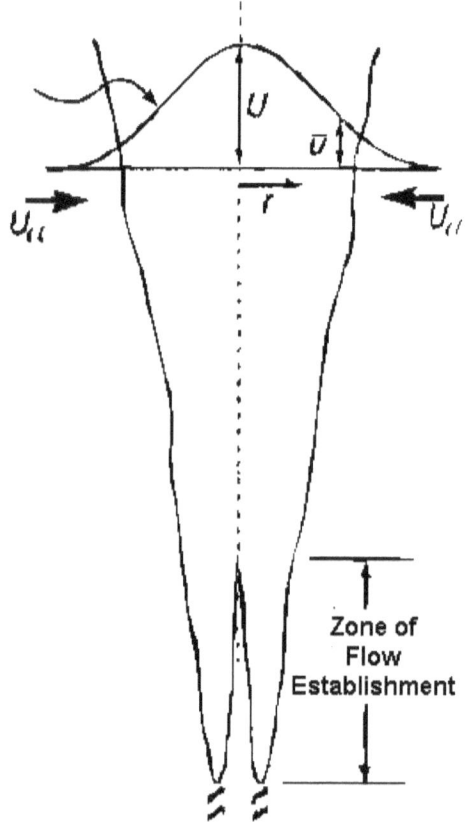

The first three assumptions/simplifications listed previously are inherent in this conceptual model. The discharge is distinct from the ambient. The ambient is assumed large in comparison to the plume. The extent of the plume is identified by the boundary, where the entrainment is assumed to occur. This figure shows the discharge from adjacent jets merging to form a single plume.

Large-Scale Turbulence and Time-Averaging

Large-scale turbulence can be clearly seen in the figure below, which is a photograph taken by Lee et al.[31] of a thermal plume revealed with blue dye. Spatial variability in plumes like the one shown are part of common experience. Large plumes caused by discharges into water bodies such as rivers and lakes are often described as *boils*, even if there is no heat involved. The discharges from some power plants such as TVA's Browns Ferry and Sequoyah Nuclear appear on the surface of the river as irregular areas with different rippling than the surrounding water, making them visually distinct. These *boils* slowly drift back and forth in the vicinity of the diffuser. The rate and extent of drift depends on the local conditions, but the movement is certainly noticeable over a period of fifteen minutes to an hour.

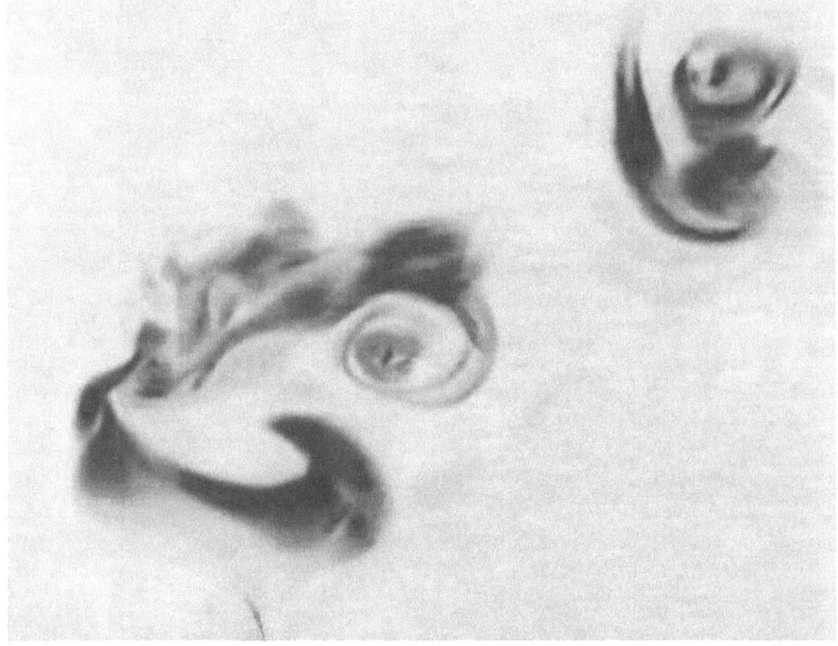

Considering this commonly observed spatial and temporal variability in plumes, it is remarkable that any agreement can be achieved between a model and field data; but two field data sets have already been introduced which have acceptable agreement. The key lies in time averaging. Combining the technologies of laser-Doppler anemometry, laser-induced fluorescence, and digital image processing, Lee et al. have been able to produce pictures, which graphically reveal this phenomenon. These next two figures show the instantaneous and time

---

[31] Lee, J. H. W., W. Rodi, and C. F. Wong, "Turbulent Line Momentum Puffs," ASCE Journal of Engineering Mechanics, Vol. 122, pp. 19 29, 1996.

averaged concentration, respectively, resulting from the discharge of a single jet as viewed from the end.

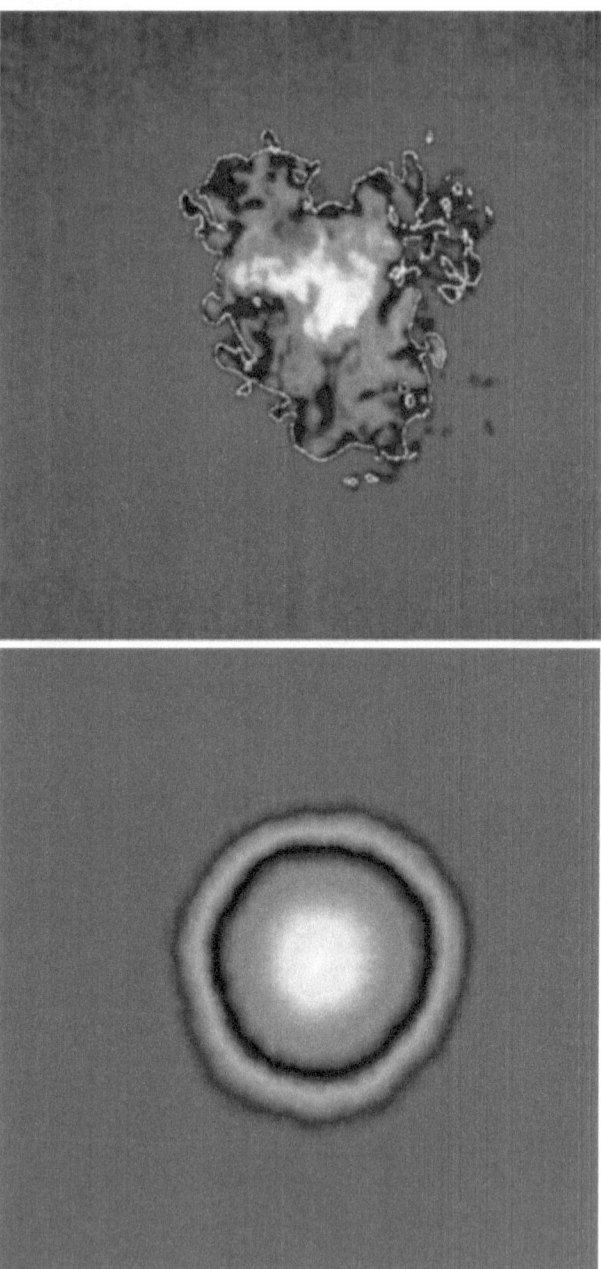

The figure below shows the instantaneous concentration of the same jet as viewed from the side.

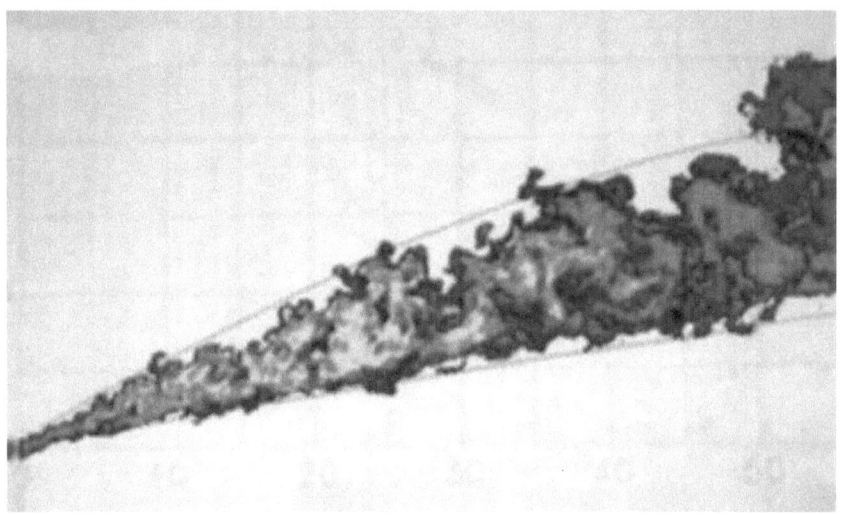

The figure below shows the time-averaged side view. The contrast between these two temporal perspectives is quite striking.

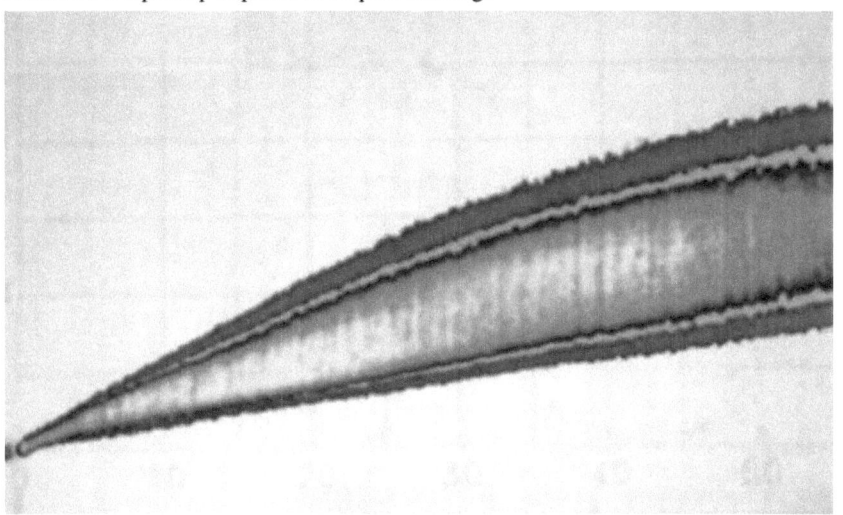

In addition to the qualitative comparison of these two figures, the next two provide a quantitative comparison. First, the instantaneous then the time averaged:

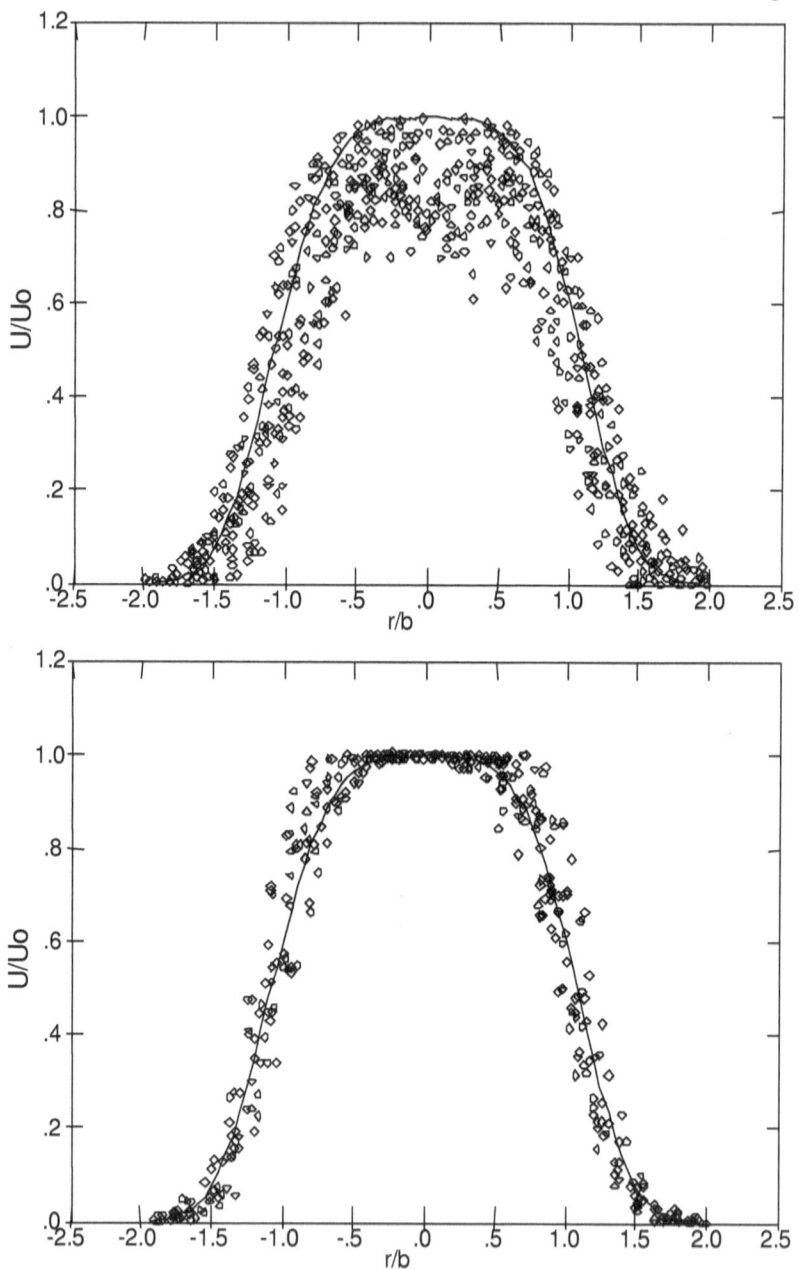

These data cover approximately the first fourth (left side) of the jet. These data are presented as normalized radius (local/total) vs. normalized concentration (local/centerline). The difference between the second pair of figures is significant, though not as dramatic as the previous pair, as they represent only the initial part of the jet. Also shown in the figures is the theoretical (Gaussian) distribution. The variance between data and theory for the instantaneous measurements is more than four times that for the time-averaged measurements. Lateral entrainment is illustrated in this next figure.

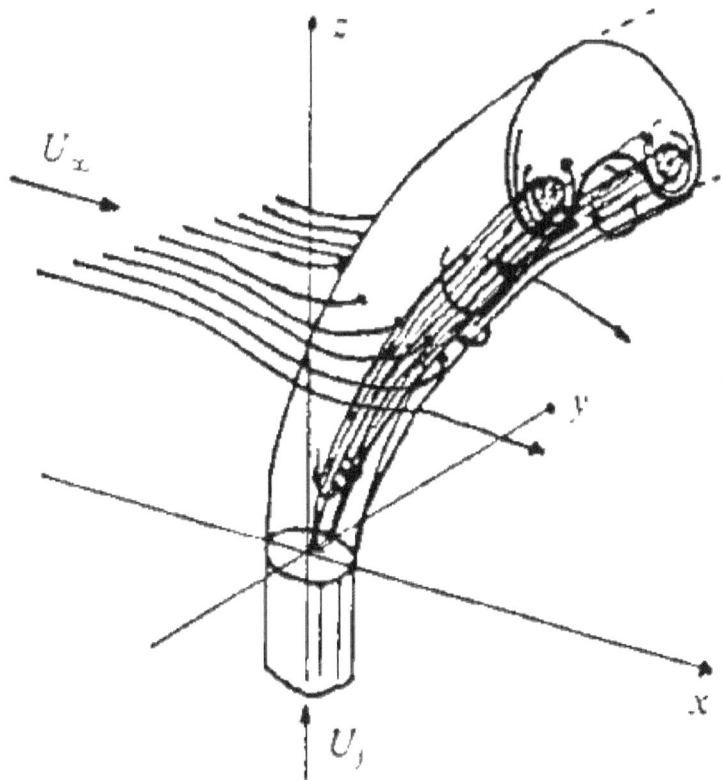

Measurements taken on an instantaneous basis are unlikely to agree locally with any model; whereas, measurements taken on the averaged time scale might be adequately predicted by a model. It appears likely from the preceding figures that instantaneous measurements taken at the edges of the plume would differ significantly from the time-averaged measurements taken in the same locations and that the discrepancy would be significantly greater near the edges than near the centerline.

Observation of plumes such as the ones at TVA's Browns Ferry and Sequoyah Nuclear Plants indicates that the time scale over which point measurements must be averaged in order to obtain values sufficiently representative of the large-scale turbulent variability is at least fifteen minutes, and may be as long as an hour. The field data already introduced are hourly averages. Consideration of the preceding figures also indicates that field measurements should concentrate on the core of the plume. Comparison of the near-centerline velocities in latter two indicates that, even in the core of the plume, field measurements averaged over a time interval less than the time-scale of the turbulence may not be sufficiently characteristic.

Quasi-Steady-State Modeling

The numerical model, as well as all the analytical and empirical models already presented, assume that the modeling time scale will be such that the turbulent processes, though certainly not steady-state, physically approach some coherent average which can be mathematically described at a level of detail illustrated by the preceding figures. For large plumes such as the ones at TVA's Browns Ferry and Sequoyah Nuclear Plants this time scale is on the order of an hour. It is assumed that any significant changes in discharge temperature or flow and ambient temperature or flow occur slowly enough that the turbulent processes have enough time to sufficiently approach this coherent average in order to be adequately described by simply averaging these values. A longer modeling period, say on the order of a day, would be treated as a sequence of quasi-steady states, thus the term, *quasi-steady-state* modeling is applicable here.

Although it is beyond the scope of this modeling to consider transients and the specifics of turbulence, it is worth noting that turbulence appears to behave randomly and is treated as a random process in three-dimensional transient computational fluid dynamics models that consider the local details of turbulence. It is highly unlikely that any three-dimensional transient model simulating random turbulence could ever quantitatively match the *swirls*, as shown previously in this chapter. Certainly, something qualitatively like the swirling dye pattern could be generated, but a point-by-point comparison of computed and measured parameters would likely be as random as the turbulence itself.

Internal Distributions

It is often assumed that a plume will have some internal distribution of velocity, temperature, or concentration relative to the centerline. Laboratory and field measurements indicate this is the case. The most common distribution is Gaussian, or an exponential decay, as this is an analytical solution to similar problems and can easily be integrated. A Gaussian distribution also fits reasonably well with data.

Consider any scalar value represented by, C. The average value along the centerline of a jet can be computed by Equation 4.29:

$$\overline{C} = \frac{\int_0^\infty C(r) e^{\left(-\frac{r^2}{b^2}\right)} 2\pi \left(\frac{r}{b}\right) \frac{dr}{b}}{\int_0^\infty e^{\left(-\frac{r^2}{b^2}\right)} 2\pi \left(\frac{r}{b}\right) \frac{dr}{b}} \qquad (4.29)$$

where r is the distance from the centerline and b is the local width of the jet. If the distributions along the centerline are similar, that is, when appropriately normalized they have the same shape, then Equation 4.26 results in a constant. For this reason it is often assumed that the ratio of the mean and maximum values of any scalar quantity is a constant over the entire plume, and so it is in this model. There is, of course, a difference between the arithmetic and integrated average values of nonlinear quantities, such as velocity squared, for which some empirical correction could be applied.

Problems arise with internal distributions when considering entrainment. If the entrainment is assumed to be proportional to the difference between the ambient velocity and the maximum velocity within the plume, then the average velocity within the plume will never equal the ambient velocity, regardless of the duration of interaction. If the entrainment is assumed to be proportional to the difference between the ambient velocity and the average velocity within the plume, then the eventual average velocity of the plume will equal the ambient. However, the maximum velocity within the plume will exceed the ambient velocity. No internal distributions are assumed in this model.

Numerical Model Assumptions

Based on the preceding arguments the following assumptions are made in developing the numerical model:

1) The discharge plume/jet is a coherent object, distinct from the ambient, which can be characterized by a similar profile or distribution and mean parameters.
2) The discharge plume/jet is small relative to the ambient such that it entrains the ambient, a process which changes its character but does not change the character of the ambient so much that the profile or distribution and mean parameters characterizing the ambient are no longer meaningful.
3) Interaction between the plume/jet and the ambient occurs at the boundary between the two and is limited to entrainment.
4) Small-scale turbulent phenomena can be characterized by bulk parameters such as mixing length and turbulent viscosity.
5) Large-scale turbulent phenomena can be averaged over time and accounted for by empirical factors such as entrainment coefficients.
6) The temporal behavior of the plume/jet is assumed to be quasi-steady-state.

7) The internal structure and properties of the plume/jet (density, velocity, temperature, salinity, etc.) can be characterized by average parameters that vary along the centerline with the trajectory.

Plume Differential Element

The computational element on which the numerical model is developed is based on these seven assumptions. The plume variables and flows are illustrated in this next figure, where m is the mass flow [kg/s], E is the energy flow [kJ/s], and C is the flow of any extensive scalar property [units/s]. The corresponding intensive properties are also shown in the figure, e is the specific energy [kJ/kg] and c is the specific scalar property [units/kg]. The subscripts A, P, and E indicate the ambient, plume, and entrainment, respectively. A Taylor series expansion is used to represent the change of variables through the element and forms the basis for generating the ordinary differential equations. Ordinary differential equations, rather than partial differential equations, arise from Assumption 7, which implies that there is only one independent variable, the distance along the centerline, s.

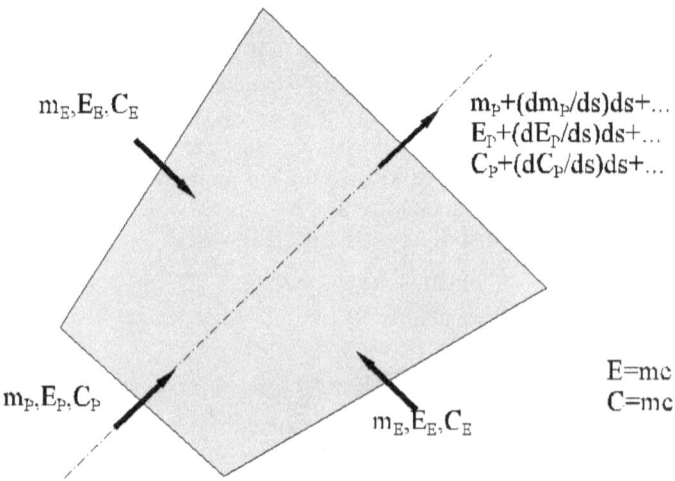

Governing Differential Equations

The applicable governing principles are: the conservation of mass, energy, and linear momentum. When there is salt in either the discharge or ambient, it is necessary to have a separate conservation equation for the salt. The conservation of mass for a slot and round jet is given in differential form by Equations 4.30 and 4.31, respectively:

$$\frac{d\left(b\,\rho_P\sqrt{u_P^2 + w_P^2}\right)}{ds} = m_E \qquad (4.30)$$

$$\frac{d\left(\pi\,r^2\,\rho_P\sqrt{u_P^2 + w_P^2}\right)}{ds} = 2\pi r\,m_E \qquad (4.31)$$

where u_P and w_P are the x and z components of the velocity along the centerline [m/s] and m_E is the entrainment mass flux [kg/m/s]. The conservation of energy for a slot and round jet is given in differential form by Equations 4.32 and 4.33, respectively:

$$\frac{d\left(b\rho_P C T_P \sqrt{u_P^2 + w_P^2}\right)}{ds} = m_E C T_E \tag{4.32}$$

$$\frac{d\left(\pi r^2 \rho_P C T_P \sqrt{u_P^2 + w_P^2}\right)}{ds} = 2\pi r\, m_E C T_E \tag{4.33}$$

where C is the specific heat [kJ/kg/°C] and T is the temperature [°C]. The conservation of linear momentum has two orthogonal vector components, x and z [m], and for a slot jet these are given in differential form by Equations 4.34 and 4.35:

$$\frac{d\left(b\rho_P u_P \sqrt{u_P^2 + w_P^2}\right)}{ds} = m_E u_E \tag{4.34}$$

$$\frac{d\left(b\rho_P w_P \sqrt{u_P^2 + w_P^2}\right)}{ds} = m_E w_E + b\left(\rho_E - \rho_P\right)g \tag{4.35}$$

where the last term in Equation 4.35 is the contribution of buoyancy. For a round jet the two orthogonal components of the conservation of linear momentum are given in differential form by Equations 4.36 and 4.37:

$$\frac{d\left(\pi r^2 \rho_P u_P \sqrt{u_P^2 + w_P^2}\right)}{ds} = 2\pi r\, m_E u_E \tag{4.36}$$

$$\frac{d\left(\pi r^2 \rho_P w_P \sqrt{u_P^2 + w_P^2}\right)}{ds} = 2\pi r\, m_E w_E + \pi r^2\left(\rho_E - \rho_P\right)g \tag{4.37}$$

The trajectory of the plume (x_P, z_P) is given in differential form by Equations 4.38 and 4.39:

$$\frac{d x_P}{ds} = u_P \tag{4.38}$$

$$\frac{d z_P}{ds} = w_P \tag{4.39}$$

The only term not already defined is the entrainment flux, m_E, which will be described after the method of solution, as any number of relationships for entrainment could be used in this formulation.

Solution of the Governing Equations

These ten ordinary differential equations express the behavior of the plume. In addition to the differential equations, it is necessary to prescribe the initial conditions and the conditions of the ambient. Once these are defined, the trajectory and dilution of the plume can be determined by solving the equations numerically. A satisfactory method is fourth order Runge-Kutta. A limited geometric progression of the step size can be used to speed the process without loss of accuracy.

Entrainment

The entrainment flux, m_E, has been computed in diverse ways by various investigators so that several relationships exist in the literature. Typically, an entrainment coefficient, α, is defined as in Equation 4.40 for slot or round jets:

$$m_E = \alpha \rho_E |\Delta v| \qquad (4.40)$$

where Δv is the magnitude of the difference between the velocity of the plume and ambient, considered as vectors. A simple and satisfactory relationship for the entrainment coefficient based on the densimetric Froude number, F_D, is given by Fischer et al. for slot and round jets and can be expressed by Equations 4.41 and 4.42, respectively:

$$\alpha = 0.0520 \, e^{\left(\frac{1.62}{Fr^{1.5}}\right)} \qquad (4.41)$$

$$\alpha = 0.0535 \, e^{\left(\frac{1.43}{Fr^2}\right)} \qquad (4.42)$$

While the entrainment at the upper and lower or upstream and downstream boundaries of the plume are no doubt different, there is insufficient experimental information to separate the two. McIntosh inferred a relationship for the entrainment coefficient from the data reported in 1983, which can be expressed by Equation 4.43 and would apply to a slot jet:

$$\alpha = \begin{cases} 0.27, & Fr < 0.75 \\ \dfrac{0.27}{Fr^{2.5}}, & 0.75 \leq Fr \leq 1 \\ 0.55, & Fr > 1 \end{cases} \qquad (4.43)$$

These relationships are shown in the figure at the top of the next page.

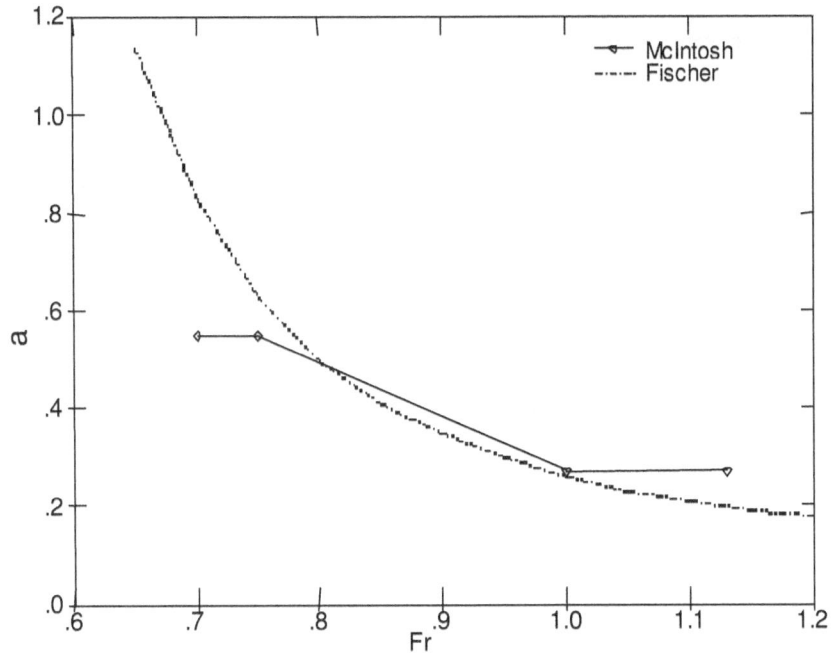

Plume Termination

Just as merging jets and real slots are handled in many different ways at the beginning of the plume, so the end of the plume is handled in various ways by different developers. The current plume termination scheme is neither the simplest nor the most complex. When the upper edge of the plume reaches the surface the entrainment on that side ceases and the vertical momentum associated with it stops. When the lower edge of the plume reaches the bottom the entrainment on that side ceases and the momentum associated with it stops. If both sides of the plume cease entraining the calculations cease. If the centerline of the plume reaches the surface or the bottom the calculations cease. After more than a fixed number of steps the calculations cease.

Plume Shape and Boundary

The primary purpose of this model is to accurately predict the entrainment or dilution. The secondary purpose is to estimate the extent of the plume, especially if it stagnates, as when a buoyant plume doesn't reach the surface or a dense plume doesn't reach the bottom. It is not the purpose of this model to accurately predict the shape or boundary of the plume. If prediction of the shape or boundary of the plume is required a three-dimensional model would be more appropriate.

Low Flow Conditions

If the flow in the river persists at a rate less than that required to completely remove the heat (or salt) an unsteady build-up will occur. Such a situation would violate the assumption of quasi-steady modeling and invalidate the results. The plume model could be used to analyze a system with feedback, such as a closed loop, but there still must be an ultimate heat (or salt) sink in order to obtain valid results.

Reverse river flow episodes can cause the effluent to travel upstream. During the subsequent forward flow episode the effluent is recycled through the plume. Recycling of the effluent through the plume is not taken into account in the development. Modeling extended periods of reverse flow is beyond the scope of this model as this would be quite speculative. Field data collection is usually avoided during reverse flow and it is not clear how a laboratory model could be constructed in order to accurately represent the prototype under such conditions.

Results

The results which led to the development of the plume model are shown in the following figure, where the red line indicates monitored values and the green line indicates computed values.

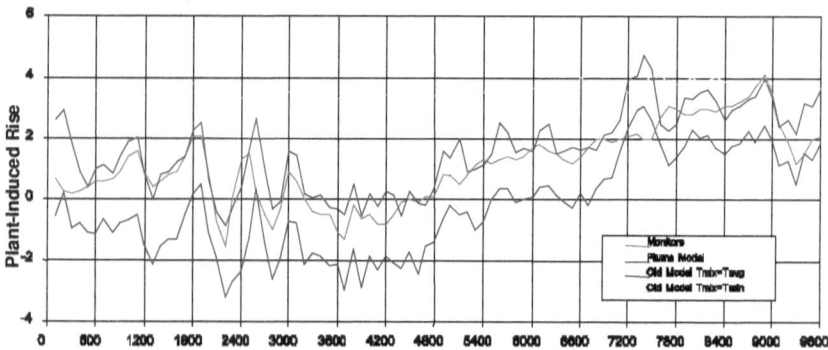

On May 5, 1981, the Sequoyah Nuclear Plant Operational Scheduling Model was predicting a steady plant-induced temperature rise in the river of 2.5°C; but the continuous monitors located in the river were reporting no more than a 1°C difference in the upstream and downstream temperatures in the compliance zone at an average depth of 1.5 meters below the surface. During the evening of May 5 and the morning of May 6 the monitored plant-induced rise fluctuated between positive and negative values. Throughout the rest of May 6 and the morning of May 7 the plant-induced rise remained fairly steady between ±0.5°C. For the rest of May 7 and all of May 8 the plant-induced rise slowly climbed to about 2°C. Throughout this entire period the plant was discharging almost 2300 MWt (megawatts-thermal) of waste heat into the river. Everything was checked, nothing was malfunctioning, but something was missing from the analysis: stratification.

Impact of Stratification

Not only are the temperatures in river upstream and downstream of the compliance zone monitored, but so is the temperature of the water being withdrawn from the river and the water being discharged back into the river as it enters the diffuser. During this period a steady flow of 35.5m³/s of water was being heated 15.5°C. By the time this heated discharge reached the surface it was often colder than the surrounding water. The vertical temperature profiles upstream of the plant are shown at six-hour intervals in the next two figures.

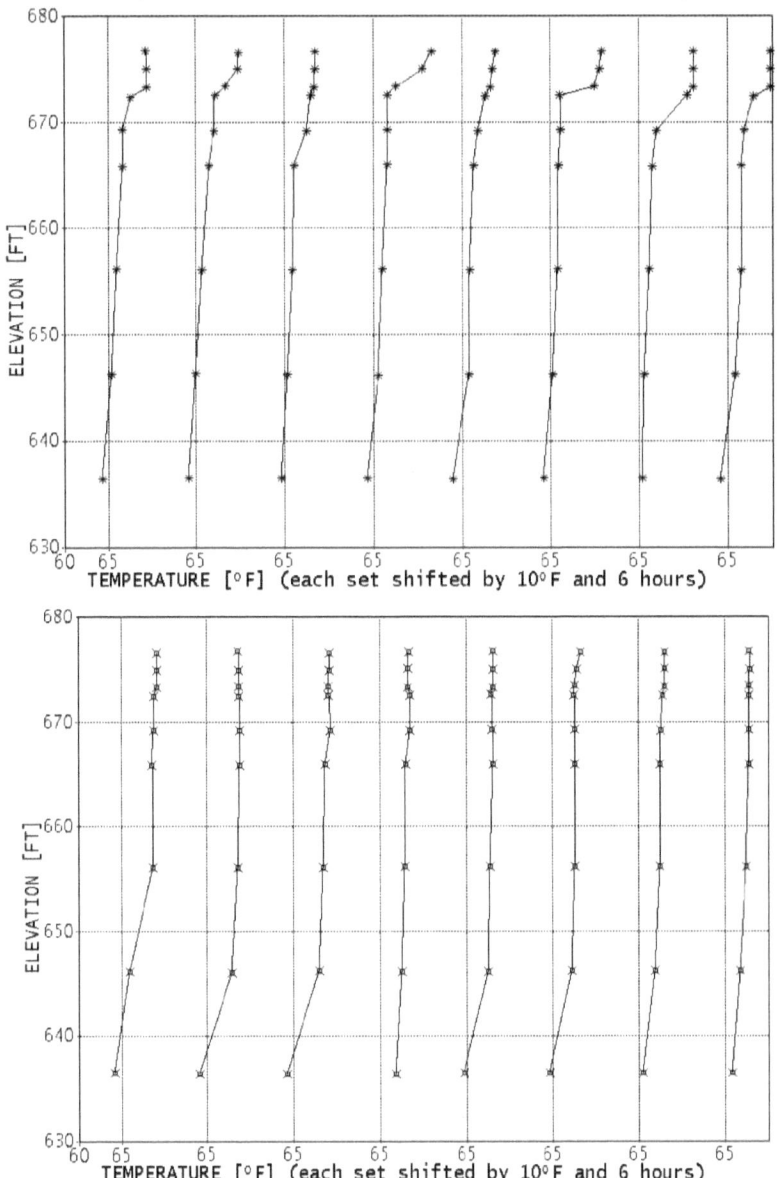

The compliance temperature is the average of the top three points, which are maintained by a float at 1.0, 1.5, and 2.0 meters below the surface. When the red line in the preceding figure passes through zero the average temperature at a depth of 1.5m below the surface upstream and downstream of the plant were the same. In which case the stratification was significant enough to absorb the entire waste heat output of the plant. When the red line in the preceding figure is below zero the stratification was more than enough. The flows during this period are shown in the figure below:

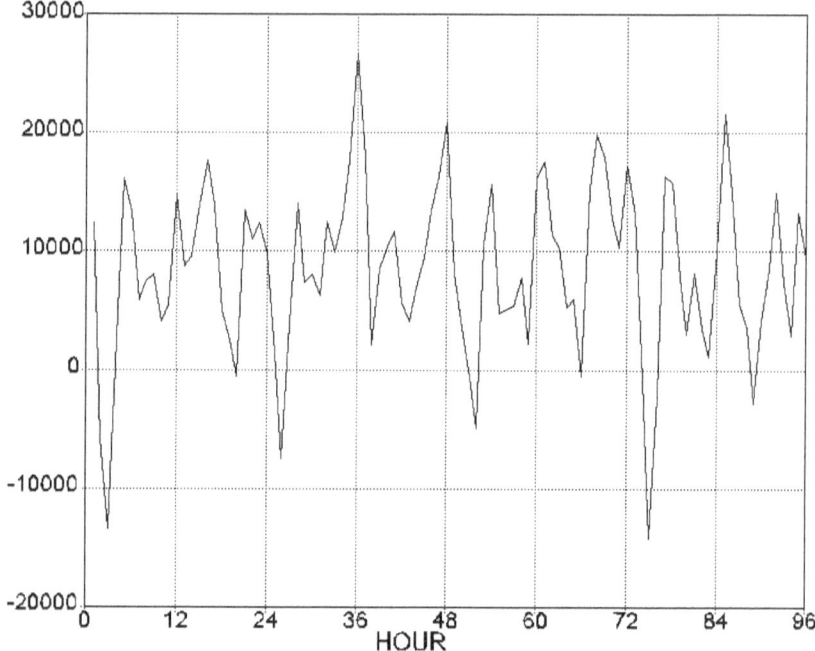

As mentioned previously several empirical models were also tested as were various temperature averaging schemes, but satisfactory agreement between model and monitors was not obtained. The best results of this effort using the empirical model of Almquist are shown by the brown line in the preceding figure.

The unadjusted results for all the empirical models tested are shown in this next figure.

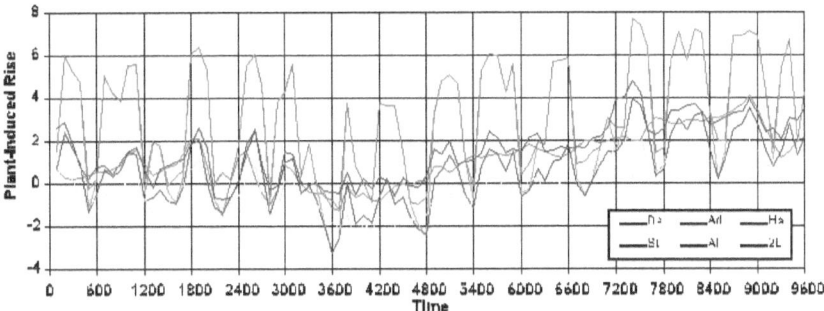

The adjusted results for the same empirical models using the best upstream temperature averaging are shown in the figure above. None of these models were developed for use in a stratified ambient.

Plume Stagnation

The analytical and empirical efforts to predict plume behavior in a stratified ambient already presented focused on how high a plume will rise before stagnating. For the purposes of computing plant-induced rise this is the most important quantity to predict accurately. If the plume does not reach the compliance zone it does not matter what the dilution is. The second most important quantity is predicting at what temperature (or salinity) is the ambient being entrained. The actual total volumetric entrainment is the least important variable in a strongly stratified ambient. This is why no upstream temperature averaging scheme will close the gap between the brown (computed) and red (measured) lines in the previous figure. Even with the exact volumetric dilution and mean entrainment temperature a plume may behave as the one at Sequoyah did from 6:00 p.m. (1800 hours in the figures) on May 5 to 6:00 a.m. (3000 hours in the figures) on May 6 when the stagnation zone drifted in and out of the monitors during essentially steady operation causing the apparent plant-induced rise to cycle from positive to negative. A plume model might predict such a response; but the empirical models tested would not.

Comparison with Monitor Data

The preceding figure shows the plume model closely tracking monitor data over a 96-hour period during which time the ambient temperature varied considerably as seen in the two previous figures. The shifting phase difference between the monitor data and plume model is a result of the distance between the upstream and downstream monitors, the varying river flow (as shown in the last figure), and the resulting travel time. The upstream ambient temperature stratification changed from gradual over the depth at 6:00 a.m. on May 5 (the left-most vertical line in the previous figure) to gradual over most of the depth with a strong gradient near the surface at 6:00 p.m. on May 6 (the second right-most vertical line in the previous figure). During the next 48 hours the upstream temperature was nearly uniform over the upper half of the river with a gradient over the lower half (the vertical lines in the same figure).

This 96-hour period provides a diverse test of the plume model with variable stratification and flows ranging from -500 to +750 m^3/s. The plant-induced rise varied from positive to negative. The plume reached the surface during most of this period but stagnated several times, never reaching the top three sensors and registering its presence. The correlation coefficients, R^2, for the models are as follows: Adams 0.14, Harleman 0.14, Stolzenbach 0.51, Almquist 0.45, the one-dimensional plume model 0.32, and the two-dimensional plume model 0.69.

Comparison with Laboratory Data

This next figure shows agreement between the laboratory data of Harleman et al. and the empirical models and the numerical plume model. The correlation coefficients, R^2, for the models are as follows: Adams 0.03, Harleman 0.03, Stolzenbach 0.28, Almquist 0.29, the one-dimensional plume model 0.27, and the two-dimensional plume model 0.48.

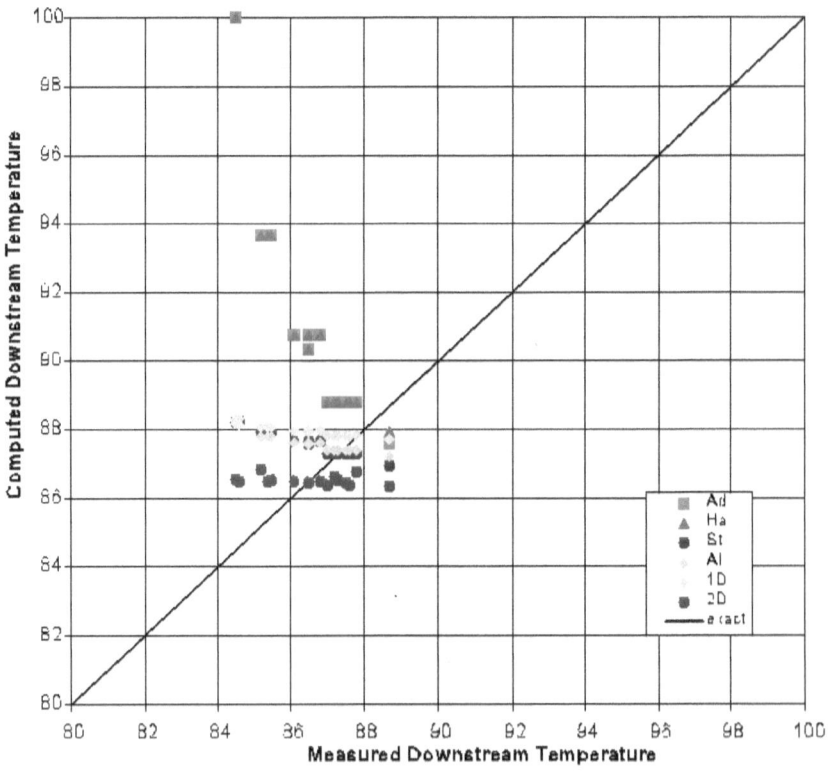

This next figure shows agreement between the laboratory data of Almquist[32] and the empirical models and the numerical plume model. The correlation coefficients, R^2, for the models are as follows: Adams 0.03, Harleman <0.01, Stolzenbach 0.55, Almquist 0.42, the one-dimensional plume model 0.12, and the two-dimensional plume model 0.82.

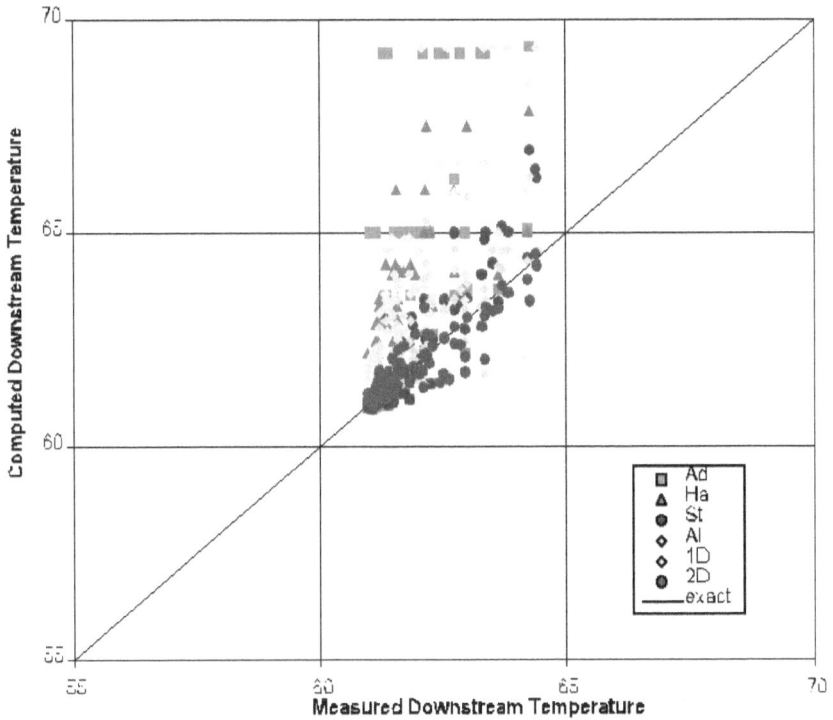

---

[32] Almquist, C. W., "Submerged Multiport Diffuser Analysis and Design for Hartsville Nuclear Plant," TVA Report WR28 1 89 100, 1978.

This next figure agreement between the laboratory data of McIntosh and the empirical models and the numerical plume model. The correlation coefficients, R^2, for the models are as follows: Adams 0.38, Harleman 0.54, Stolzenbach 0.82, Almquist 0.62, the one-dimensional plume model 0.74, and the two-dimensional plume model 0.81.

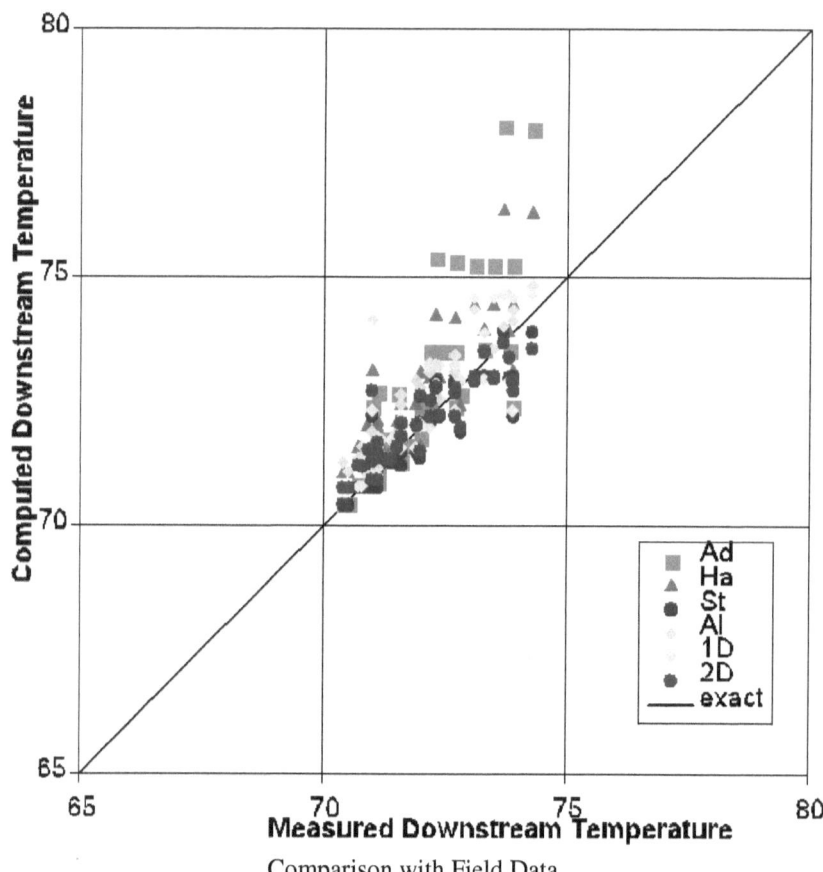

Comparison with Field Data

The figure on the top of the next page shows agreement between the field data of Almquist et al. and the empirical models and the numerical plume model. The correlation coefficients, R^2, for the models are as follows: Adams 0.04, Harleman 0.01, Stolzenbach 0.46, Almquist 0.54, the one-dimensional plume model 0.49, the two-dimensional plume model 0.60, and the three-dimensional finite-difference fluid dynamics model, EFDC, 0.51. The three-dimensional model was applied to this field data set as part of another project.

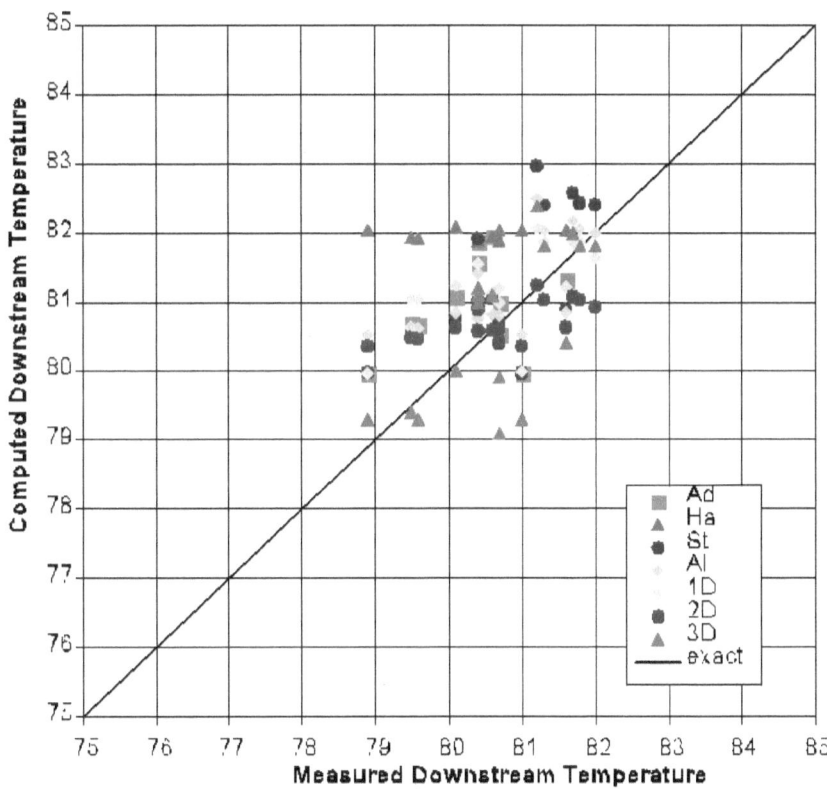

The figure on the next page shows agreement between the field data of McIntosh et al. and the empirical models and the numerical plume model. The correlation coefficients, R^2, for the models are as follows: Adams 0.91, Harleman 0.88, Stolzenbach 0.99, Almquist 0.98, the one-dimensional plume model 0.91, and the two-dimensional plume model 0.93.

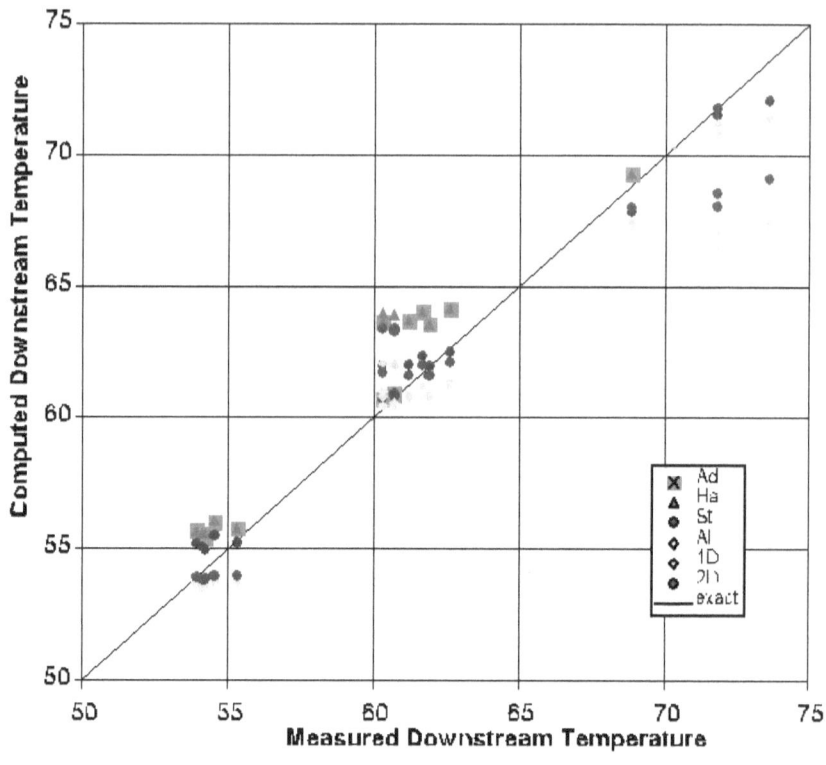

Asymptotic Behavior

The next three figures show the asymptotic behavior of the computed dilution of the plume for Froude numbers of 1, 10, and 100 respectively. Here a thermal plume is discharged upward into an infinite, uniform, stagnant ambient. Also illustrated in these figures are the results of three analytical methods already presented (Almquist, Cederwall, and Roberts) and the numerical model, SLOTJET1, by Alavian et al.[33]. The plume results agree well with Cederwall and lie between the results of Almquist and Roberts.[34]

---

[33] Alavian, V., P. Ostrowski, and J. A. Parsly, "Two Computer Models for Diffuser Performance Evaluation," TVA Report WR28 1 900 162, 1988.

[34] Roberts, P. J. W., "Line Plume and Ocean Outfall Dispersion," ASCE Journal of Hydraulics, 1979.

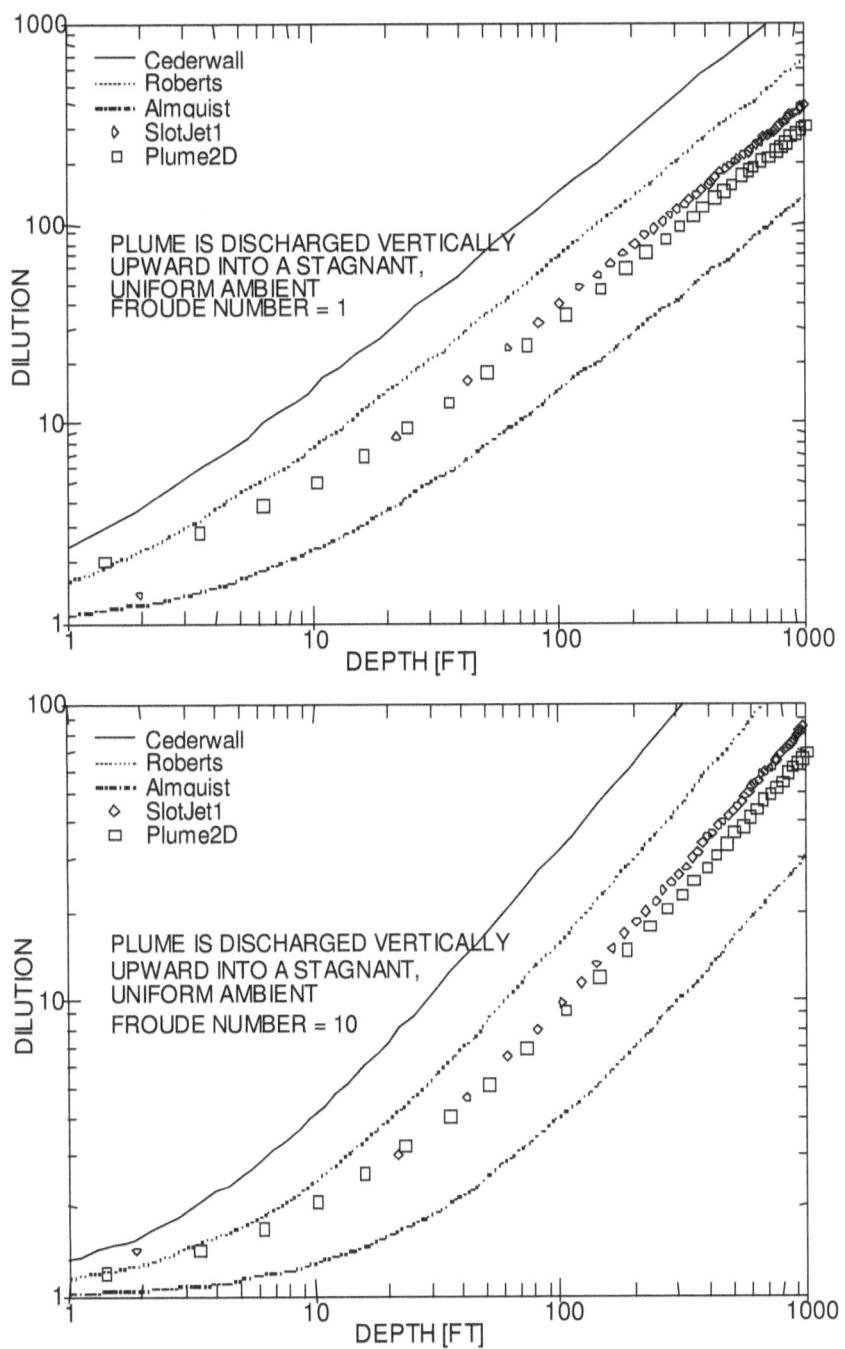

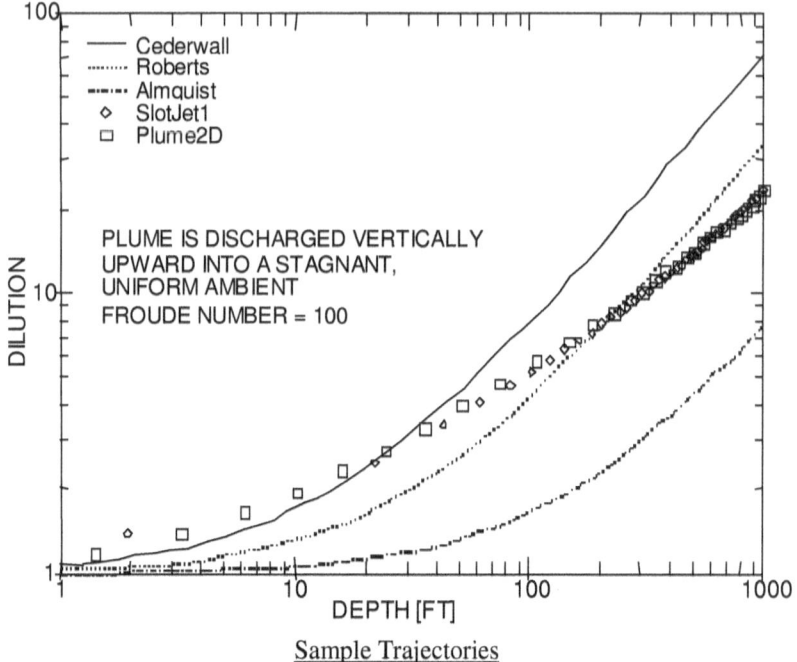

Sample Trajectories

The next two figures show the computed trajectory of the plume centerline for a range of ambient river flows. The first is for a positively buoyant thermal plume and the second is for a negatively buoyant salty plume.

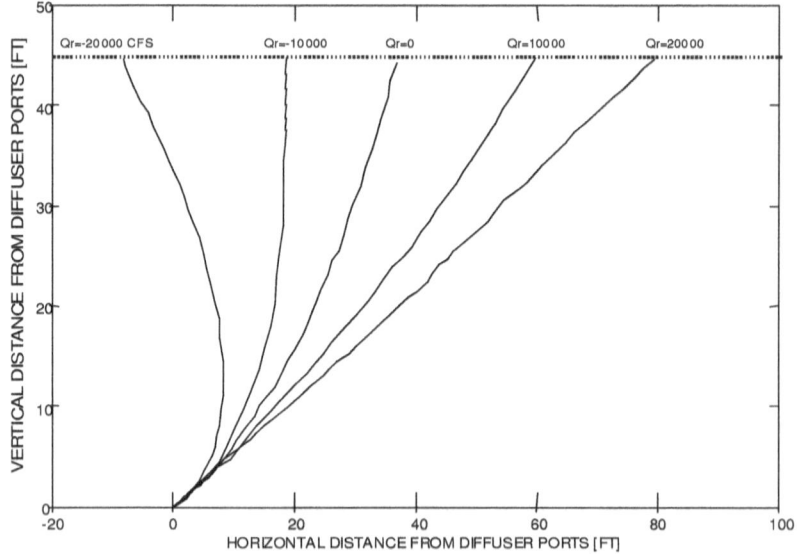

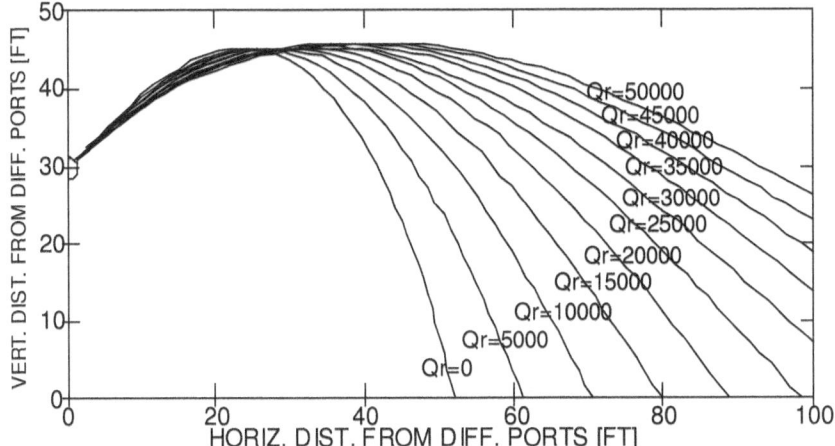

The development of a two-dimensional plume model for a round or slot jet has been presented. The formulation of this model is based on the conservative form of the governing equations. The model can handle combinations of heated and salty discharges and flowing ambients. The model results compare favorably with laboratory and field data. The model is consistent with several analytical models for the asymptotic behavior a thermal plume in an infinite medium. The model closely tracks monitor data for variable flow, ambient temperature, and upstream thermal stratification. Additional details can be found in the TVA and TWRA reports.[35,3]

Computer Code

The two-dimensional numerical plume model has been implemented in the form of a computer code. You can find the source (plume2D.c) and all of the associated input and output files in the online archive in folder examples\plume2d. Some input data are necessary to drive the model, including: the composition of the plume and the ambient temperature, salinity, and velocity. The plume may be discharged from a slot or a round jet and the equations must handle both cases. Output files are created to facilitate displaying the results graphically. Sample input files include: plume1.inp, plume2.inp, plume3.inp, and plume4.inp.

---

[35] Benton, D. J., "Development of a Two-Dimensional Plume Model for Positively and Negatively Buoyant Discharges into a Stratified Flowing Ambient," TVA Report WR28-1-45-105, 1986.

The variables are listed in the following table:

variable	units	description
α	-	entrainment coefficient
b	ft	width/diameter of the plume
depth	ft	depth
difd	ft	diffuser diameter
difl	ft	diffuser length
dify	ft	diffuser elevation (up from bottom)
dil	-	mixing ratio (dilution)
Froude	-	densimetric Froude number
Qr	ft³/sec	river flow
Qdis	ft³/sec	discharge flow
ρ	lbm/ft³	density
S	-	salinity
T	°F	temperature
ang	°	angle of inclination from the horizontal
u	ft/s	horizontal velocity of the plume
v	ft/s	vertical velocity of the plume
w	ft/s	velocity along centerline w=sqrt(u²+v²)
x	ft	horizontal distance from diffuser ports
y	ft	vertical distance from diffuser ports
z	ft	distance along the centerline

The governing equations in symbolic form are:

p(1)=x	horizontal coordinate of centerline
p(2)=y	vertical coordinate of centerline
p(3)=r*w*b²*π/4	round jet mass flux
p(3)=r*w*b	slot jet mass flux/unit diffuser length
p(4)=p(3)*u	horizontal momentum flux
p(5)=p(3)*v	vertical momentum flux
p(6)=p(3)*t	thermal energy flux
p(7)=p(3)*s	salt flux

Results for the first example (a rising plume1.inp) are shown in this figure:

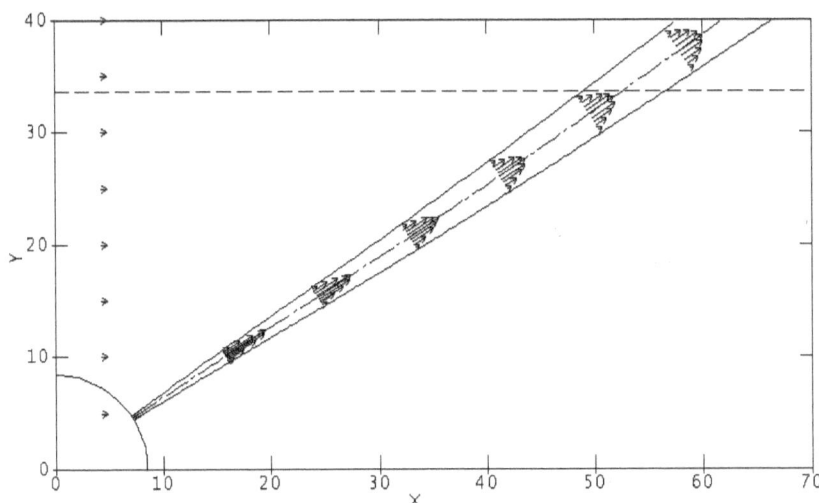

Results for the fourth example (a sinking plume4.inp) are shown in this next figure:

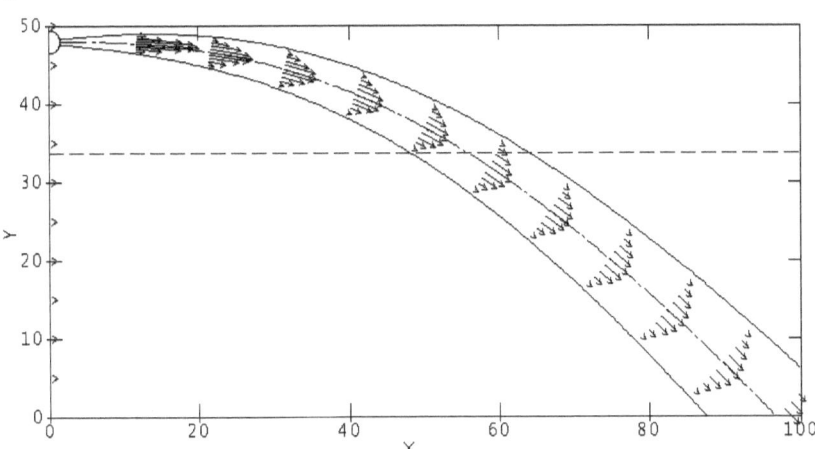

Chapter 5. 3D Thermal Plume

In the preceding chapter we only considered the receiving water on a gross basis. We will now take a much more detailed, three-dimensional approach. We first consider the geometry: three long pipes big enough to drive a truck through if they weren't under water along the bottom of a river. The bathymetry is shown in this first figure:

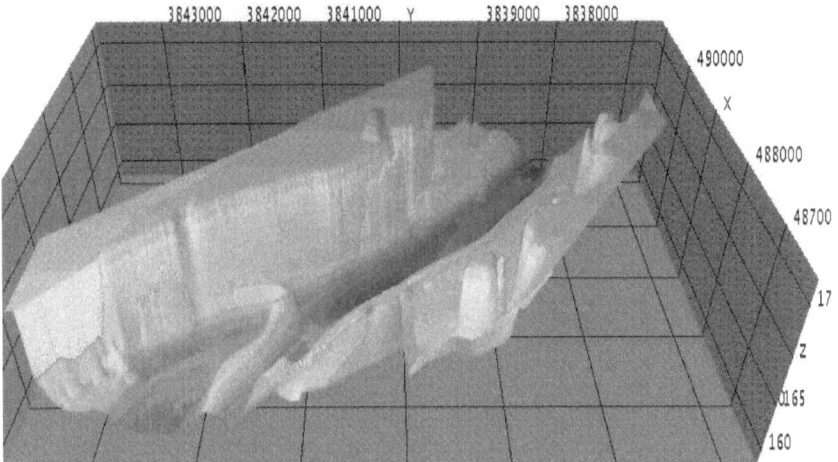

The diffusers are 20.5 ft (6.25 m), 19.0 ft (5.79 m), and 17.0 ft (5.18 m) in diameter. The lengths are 1010 ft (307.8 m), 1610 ft (490.7 m), and 2210 ft (673.6 m), respectively. Shown to scale, the pipes appear quite long and thin:

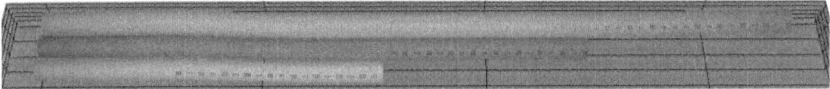

Without vertical exaggeration, it's difficult to represent the diffusers—even as large as they are—when compared to the river. The modeling is done at 1:1:1 scale for X:Y:Z.

The discharge flows are approximately 1450 ft³/s (41 m³/s) per diffuser. Water is discharged through 2-inch (0.05 m) ports spaced 6 inches (0.15 m) apart over approximately 600 feet (183 m), or about 7,000 ports per diffuser. The ports are drilled so as to span approximately 24° to 45° angle with respect to the horizontal. The navigation channel where the diffusers are located is about 850 ft (260 m) wide at this point, although the river banks are significantly farther apart than this. The diffusers are embedded in gravel and held in place by several concrete pours.

The vertical and horizontal aspect ratios are shown in the lower left corner of this next figure:

The area is shown in the following figure, including diffuser zone (in green), navigation channel, and overbank regions.

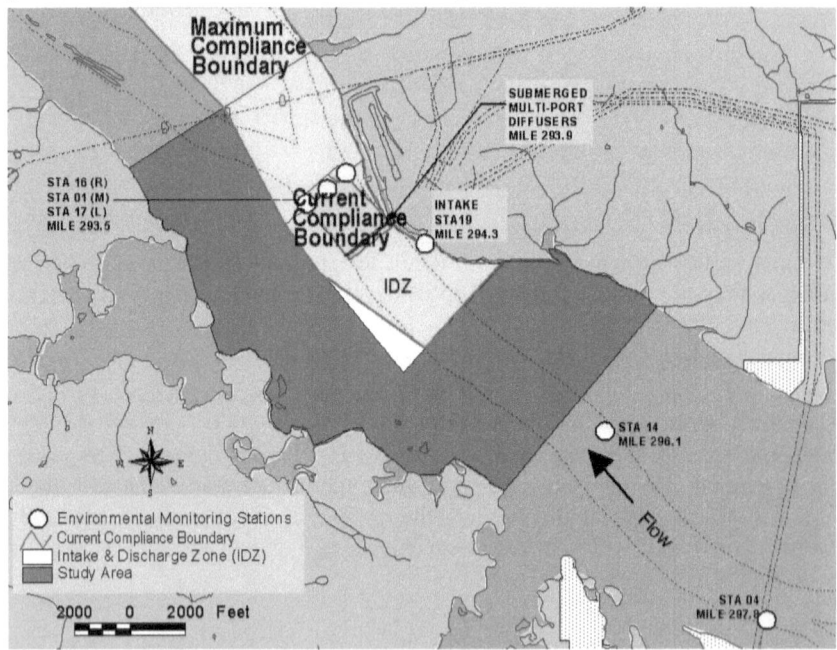

The three-dimensional flow field, including discharge from the diffusers is calculated using the Environmental Fluid Dynamics Code (EFDC) is a public domain, an open source, surface water modeling system, which includes hydrodynamic, sediment and contaminant, and water quality modules fully integrated in a single source code implementation. EFDC has been applied to over 100 water bodies including rivers, lakes, reservoirs, wetlands, estuaries, and coastal ocean regions in support of environmental assessment and management and regulatory requirements.

EFDC was originally developed at the Virginia Institute of Marine Science (VIMS) and School of Marine Science of The College of William and Mary, by Dr. John M. Hamrick beginning in 1988. This activity was supported by the Commonwealth of Virginia through a special legislative research initiative. Further developments were made by Dr. Robert Byrne, the late Dr. Bruce Neilson, and Dr. Albert Kuo, of VIMS. Subsequent support for EFDC development at VIMS was provided by the U.S. Environmental Protection Agency (EPA) and the National Oceanic and the Atmospheric Administration (NOAA) Sea Grant Program. EFDC can be obtained from the EPA at:

https://www.epa.gov/ceam/environment-fluid-dynamics-code-efdc-download-page

The figure below shows the thermal plume in cross-section. Upstream is to the right and downstream is to the left. The magenta grid indicates the thermal compliance or designated *mixing* zone. This is at low flow, which is why the thermal plume can be seen extending upstream approximately as far as it does downstream. The picture for high flow is not that much different, as shown in the next figure. The vertical exaggeration in both figures is 50:1 (Z:X/Y). This figure also shows the primary grid in black. There is also a secondary (finer) grid surrounding the diffusers in order to capture the near field effects of the jets.

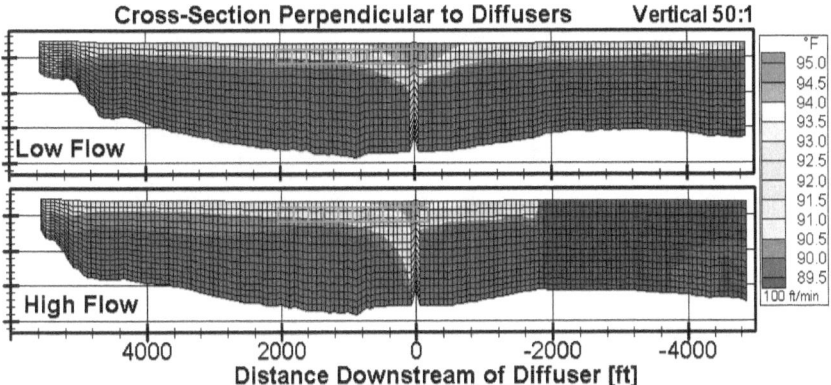

While the discharge is approximately 120°F, the zone where this temperature persists is so small that it can't be seen at this resolution. The extent of the center thermal plume is indicated by the red mesh in the following figure, which shows velocity vectors as tiny blue arrows and the channel bottom as an undulating brown surface:

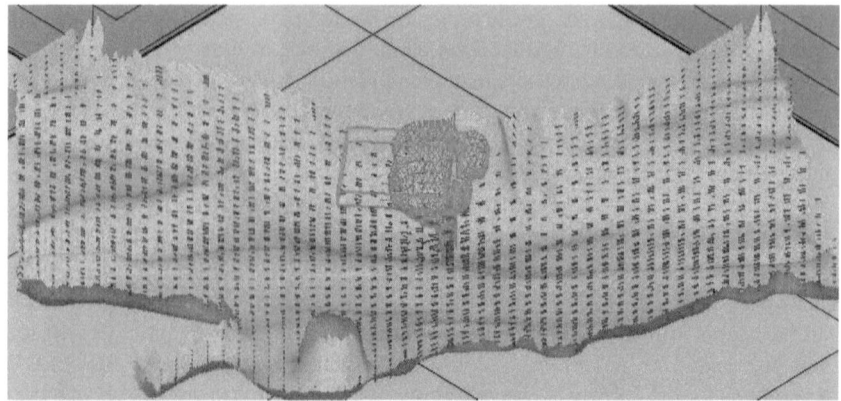

The combined plume from all three diffusers along with the compliance boundary is shown in this next figure in plan view:

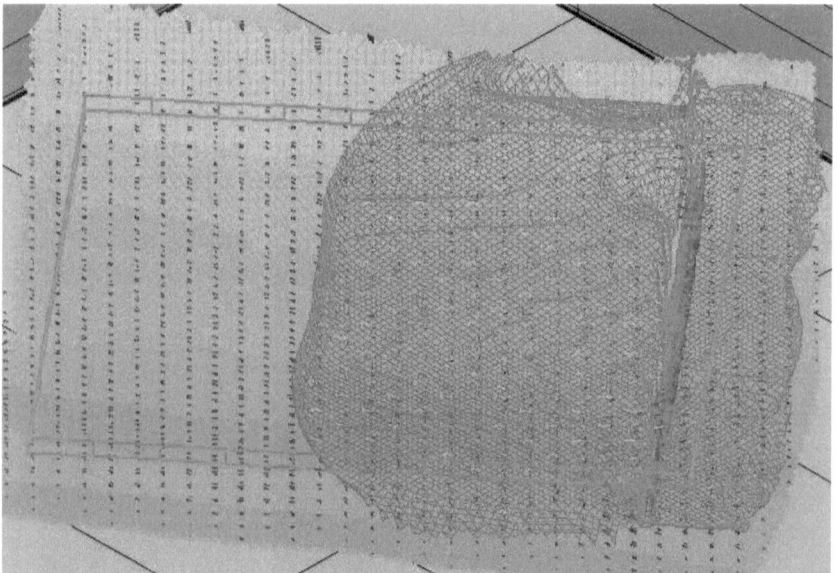

The figure below is the same model results, only at a different angle showing depth:

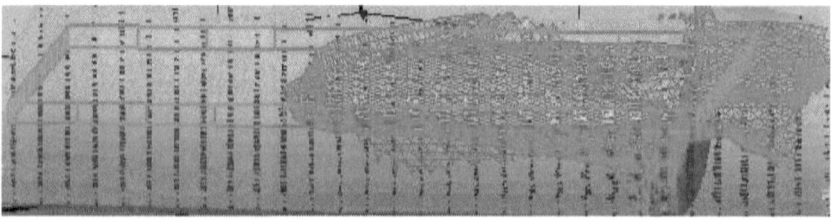

Another plan view with X:Z exaggeration of 5.7:1 and a Y:Z exaggeration of 2.5:1 is shown below. This view is looking upstream along the center of the navigation channel. The gravel and backset is also shown.

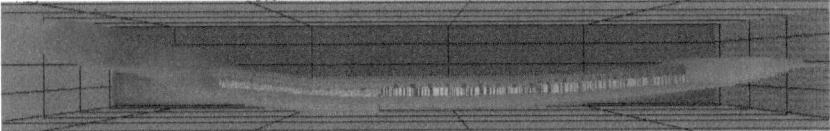

The extent of plume is shown as a blue shell and orientation of the diffusers is shown in this next figure:

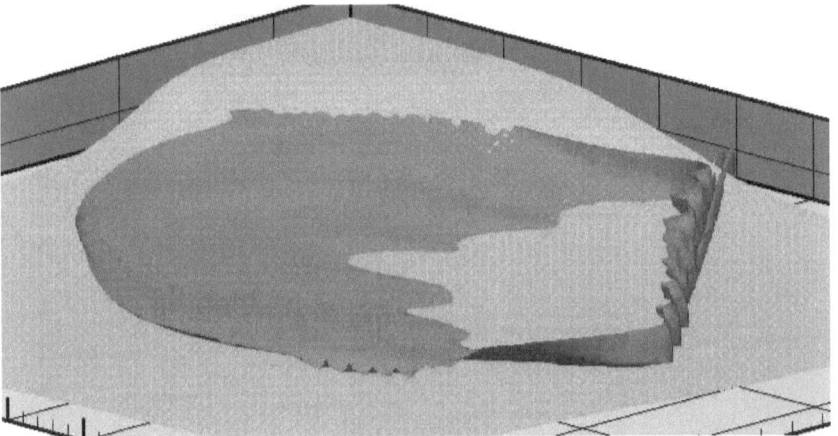

Field Data

One of the extraordinary features of this project is that the modeling effort was supported by considerable field data, which was very costly to obtain, requiring many days effort by an experienced team. The low flow data are shown in this next figure:

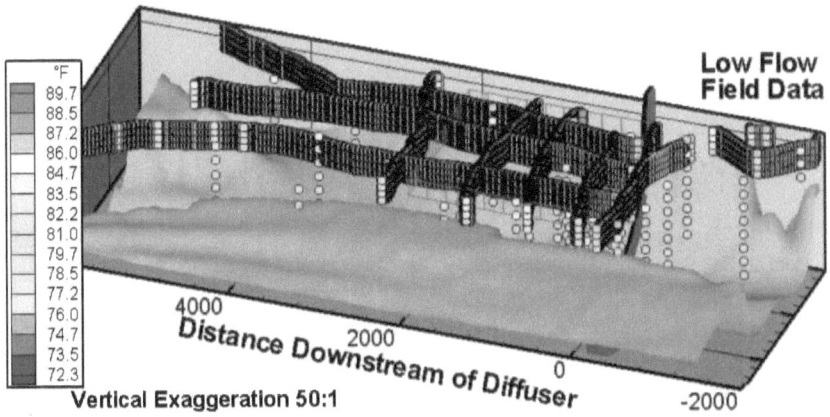

and the high flow data are shown in this figure:

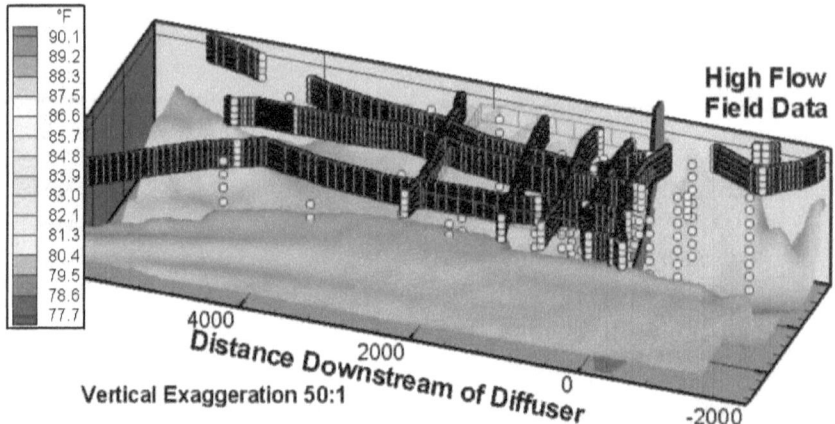

Agreement between field data for several flows, analytical models from the previous chapter, and the current 3D model are shown in this next figure:

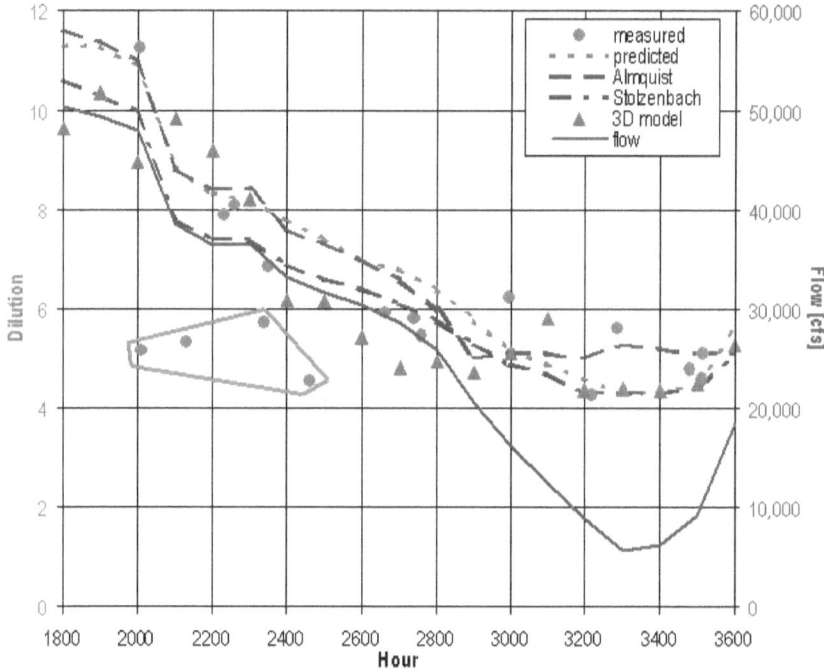

Agreement is acceptable (the green triangles are the near red circles) for high flow (left side of figure) and low flow (right side of figure) flows, except for the four data points inside the magenta box. The flow varied continuously over the 18-hour period, which is typical operation for this reservoir. While the previous analytical models (Almquist and Stolzenbach) and empirical calculation (dotted cyan curve, based on regression by Harper) and current 3D model all yield similar results for dilution, only the 3D model provided spatial details, which were of particular interest. The models of Almquist and Stolzenbach were based on laboratory (i.e., scale) models and not field data. The 3D model results were in general agreement with the field data. Graphical comparison of the field data and 3D model results were challenging, as these spanned several length scales and the field data were necessarily collected over several hours. The following graph shows the low flow data and model results:

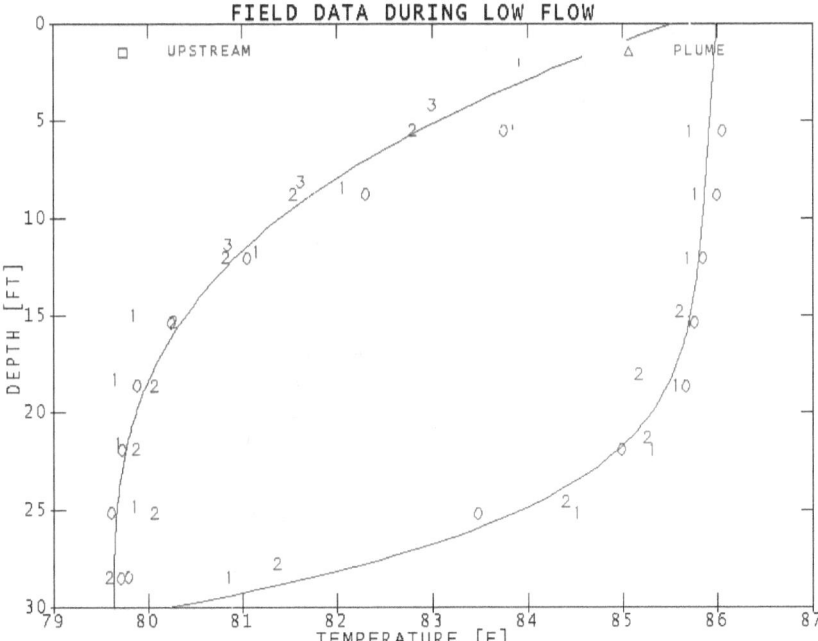

Measurements near the plume (red numbers) agree quite well with the 3D model (red curve). Upstream measurements (blue numbers) agree quite well with the 3D model boundary conditions (blue curve). Downstream measurements (green numbers) are scattered about the 3D model results (green curve), due to large-scale eddies and other transient behavior. This is not surprising, considering the complexity of this system and the duration of the data collection (several hours). Recall the instantaneous and time-averaged plumes on page 31.

Field data and 3D model results for the high flow period are shown in this next figure:

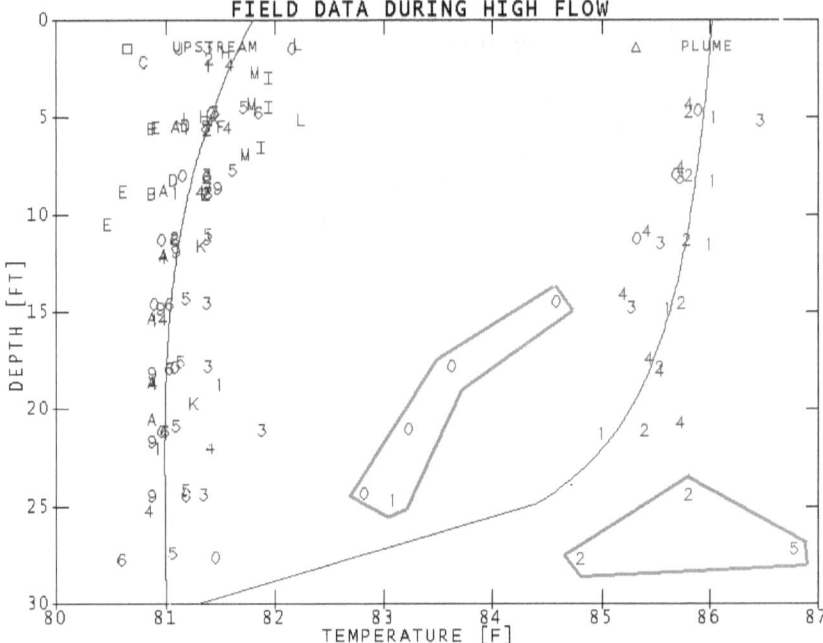

Upstream (blue letters and curve) and downstream (green numbers and curve) comparisons are similar to the low flow case, but we see more scatter in the near-plume results (red numbers and curve), as indicated by the data points in the two magenta boxes. Other than these 8 data points, the agreement is quite acceptable. Again, this is attributed to eddies and the data were collected over a span of time (18+ hours), while the 3D model runs are instantaneous snapshots.

Chapter 6. Three-Dimensional Geological Data

Most of the plumes I have modeled over the past thirty years have been contaminants in groundwater. I'm an engineer, not a geohydrologist. I first became involved with groundwater plumes by serving in the role of applied mathematician to assist a team of geohydrologists on the project known as MADE (MAcro Dispersion Experiment), which was conducted at the Columbus Air Force Base in Mississippi. Quite a few publications are available on this site, many of which are available on Research Gate.[36,37,38,39]

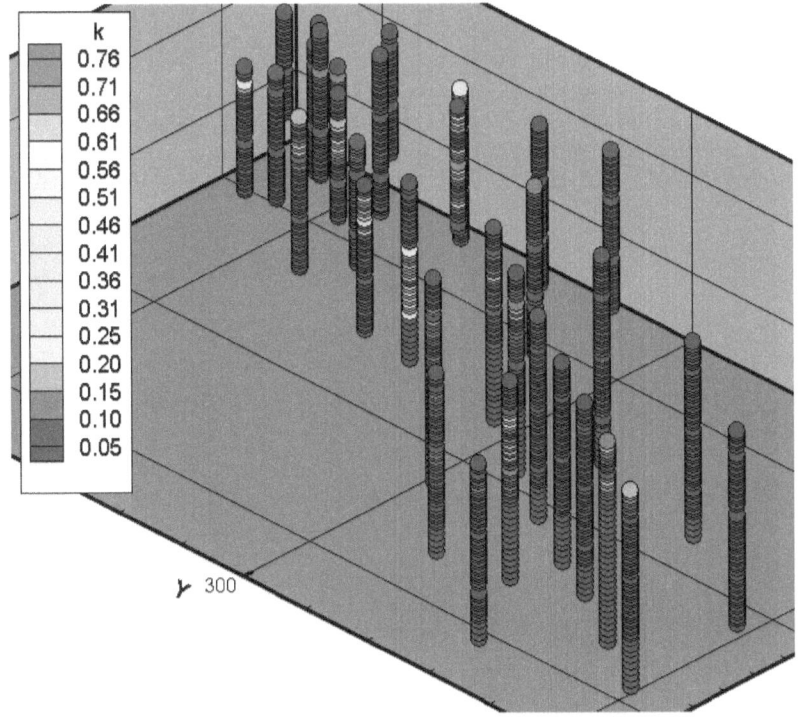

---

[36] Gelhar, L. W. and Axness, C. L, "Three-Dimensional Stochastic Analysis of Macro-Dispersion in Aquifers," Water Resources Research, January 1983.

[37] Stauffer, T. B, Boggs, J. M., and MacIntyre, W. G., "Ten Years of Research in Groundwater Transport Studies at Columbus Air Force Base, Mississippi," Biotechnology in the Sustainable Environment, December 1996.

[38] Li, L., Zhou, H., and Gomez-Hernandez, J., "A Comparative Study of Three-Dimensional Hydraulic Conductivity Upscaling at the MAcro-Dispersion Experiment (MADE) site, Columbus Air Force Base, Mississippi (USA), Journal of Hydrology, Vol. 404, No. 3, pp. 278-293, June 2011.

[39] Elfeki, A. M., and Rajabiani, "Simulation of Plume Behaviour at the Macro-Dispersion Experiment (MADE1) Site by Applying the Coupled Markov Chain Model," March 2013.

This important experiment is one of the very few large-scale groundwater experiments where tracers were intentionally injected into the ground. It is also important that work has continued at the site for so many years. Data collected at the site are uncharacteristically *good* and detailed compared to the many more unintentional contaminant sites we will cover subsequently. First, what do the data look like? For one thing, measurements come from wells. These first three figures are the first data set ever extracted from the MADE site. On the previous page we see a 3D view. This next view is in the horizontal (XY) plane.

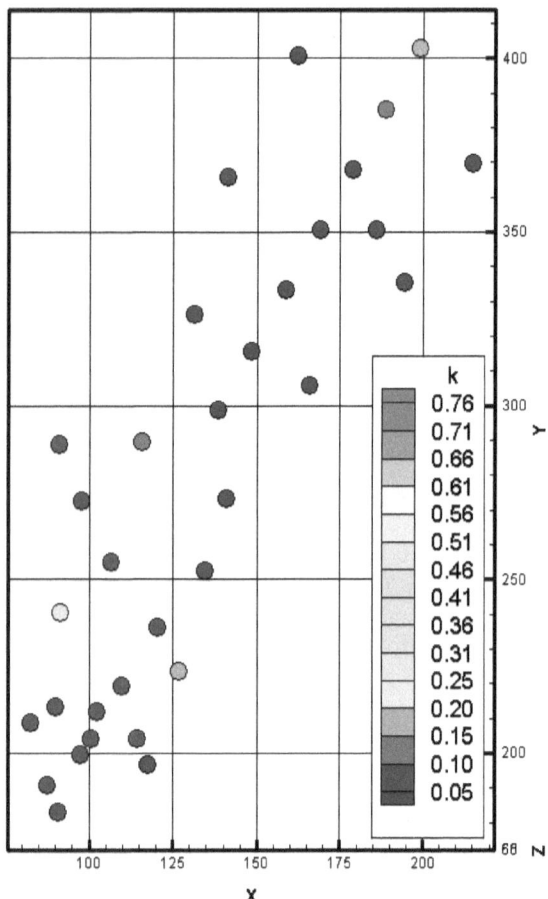

As you can clearly see, the data are not evenly spaced, which means that simplistic interpolation methods are not likely to be effective. We have already discussed the inverse distance method (Appendix A). This is just one of the methods tested on this data set.

This next view is in the vertical (YZ) plane:

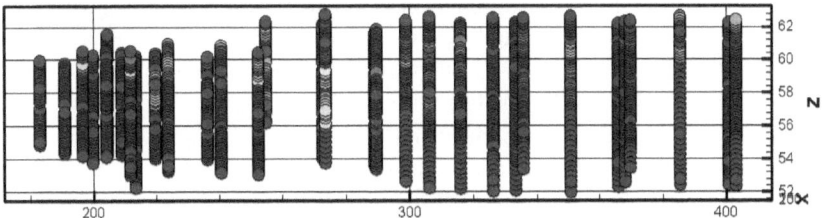

Note that this data is hydraulic conductivity and non-zero, in spite of the reported values. Reasonable bounds were placed on the data in order to obtain this representation. A bounding polygon was also applied. A 3D slicing of the interpolated field is shown below:

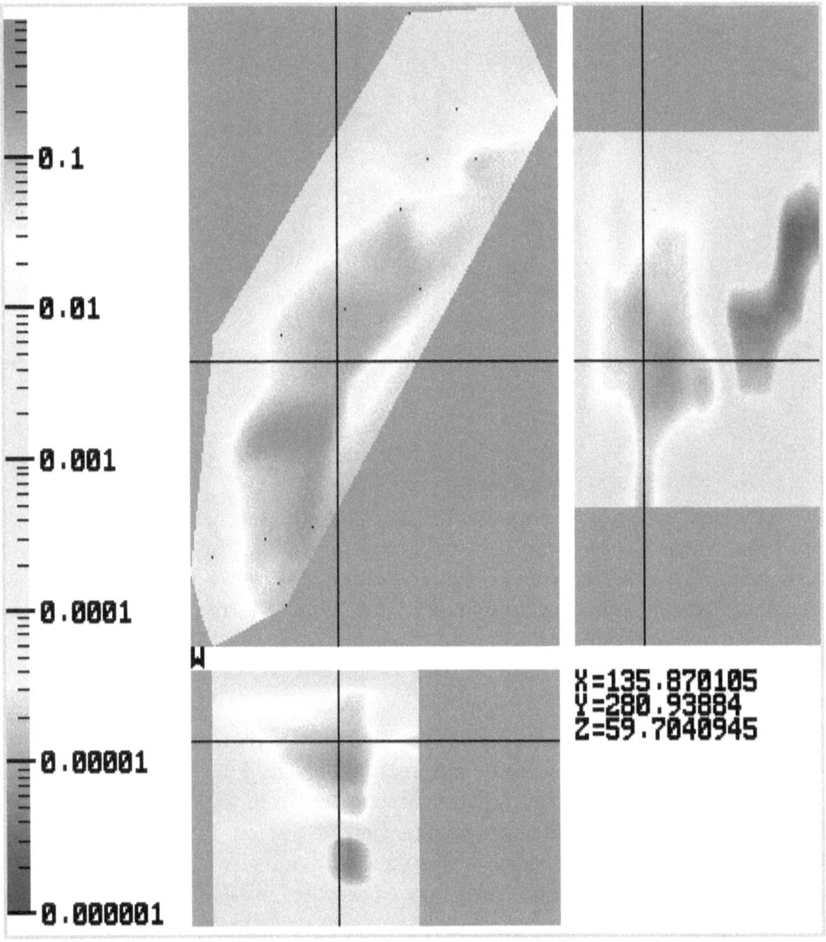

The preceding result was obtained by first applying the inverse distance method with several control points, then smoothing the results repeatedly until the spatial artifacts of well sampling were not discernable (see Appendix B for more details). This is, of course, a subjective process, which was guided by the lead hydrologist and implemented by this author.

The second data set to come from the MADE site was also hydraulic conductivity but in a different area. This came as a surprise because the values were so different, which led to a revaluation of the methods and instruments. This data was also bounded in extent. The bounding polygon is shown in the figure below as white dots:

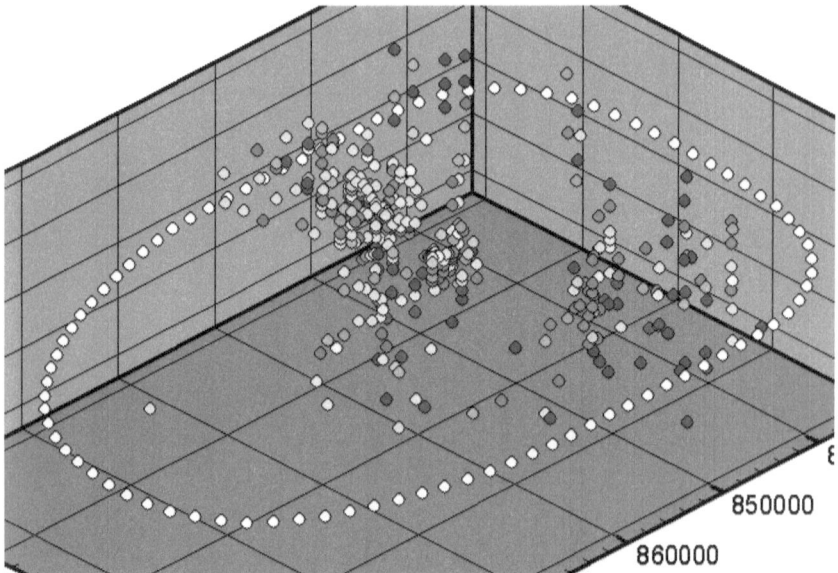

This data set proved more challenging than the first, which let the team to try other methods of analysis, including relaxation, finally settling on a combination of inverse distance and relaxation. When we get to the point of analyzing tracer concentration levels, we will use the same techniques developed for analyzing the hydraulic conductivity. This step also paved the way for numerically modeling the site, which was one of the primary goals of the study.

The second data set was so *blotchy* that it defied smooth approximation. The team eventually adopted a different approach. First, we will consider inverse distance interpolation. The results are shown in the figure at the top of the next page. This wasn't what we expected or wanted, but it does generally follow the trend that is suggested by the data. Still, it doesn't look like the sort of conductivity field you want to feed into a computer model. This was produced using Tecplot™. TP2 doesn't produce exactly the same results for nominally the same method (viz., inverse distance with octants and 2.5 power) because we modified the approach.

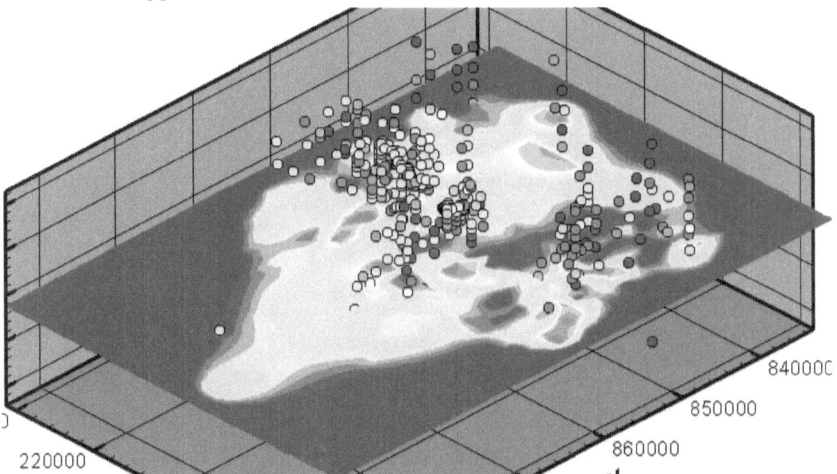

This next figure is what Tecplot™ produces using kriging:

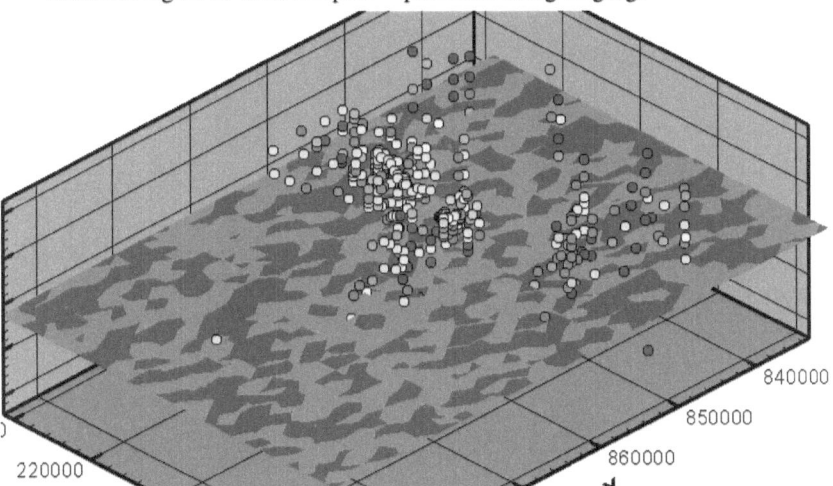

This second figure is completely unreasonable—again, no criticism of Tecplot™, which faithfully implements the conventional formulas. Up until this point, the conventional wisdom was, "No problem. Just use kriging." Clearly, the conventional wisdom is not adequate, nor is traditional kriging. Inverse distance coupled with relaxation, was built into TP2 and the project moved forward toward success. See Appendices A, B, and C for more details.

Modified kriging (see Appendix C) yields the following results, which are adequate:

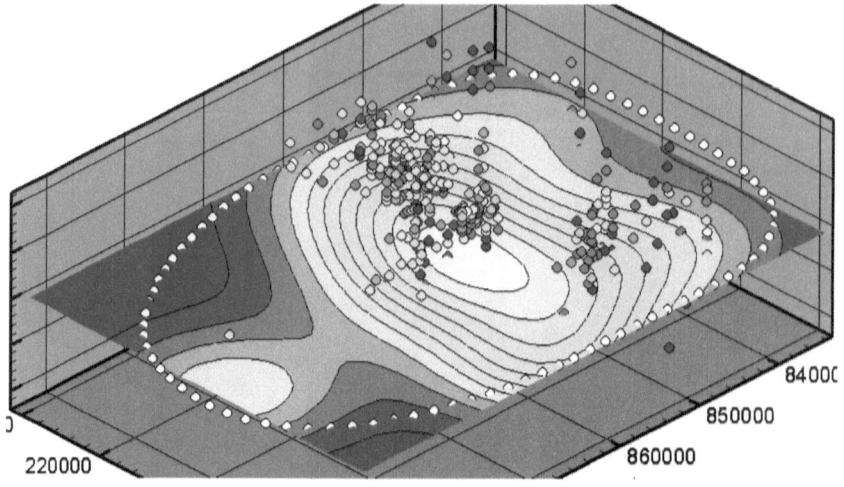

Chapter 7. Contaminant Plumes in Groundwater

This topic is where most of my efforts have been devoted and for which I have the most data. Data comes in the form of point concentrations obtained from extraction wells. This represents a snapshot in time over the duration of sampling, which is typically much shorter than the residence time or the lifetime of the plume. Data analysis techniques (and challenges) are the same as discussed in the preceding chapter.

The following figure shows the contaminant concentrations what was by far the most plentiful data set we ever worked with (15,723 points). The extent of plume is indicated by the thick magenta polygon.

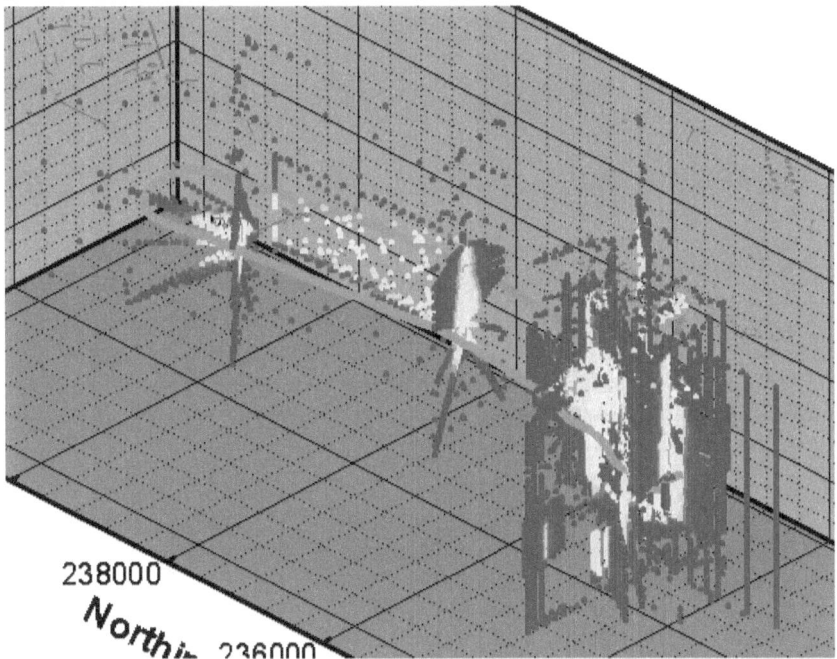

Both Tecplot™ and TP2 will extract iso-surfaces from a volume. Iso-surfaces are shells of constant value, in this case, concentration. These are quite useful for visualization. TP2 will also extract iso-lines (i.e., contours) from a surface. Iso-surfaces show the contaminant as a blob in space, which can be rotated, sliced, and viewed. Iso-surfaces are an important part of devising effective remediation strategies, especially when there are attendant structures, such as a stream, lake, or road.

Concentration iso-surfaces for the preceding data (log(C)=-5, -4, -3, -2, -1, 0, and 1) are shown in the figure below:

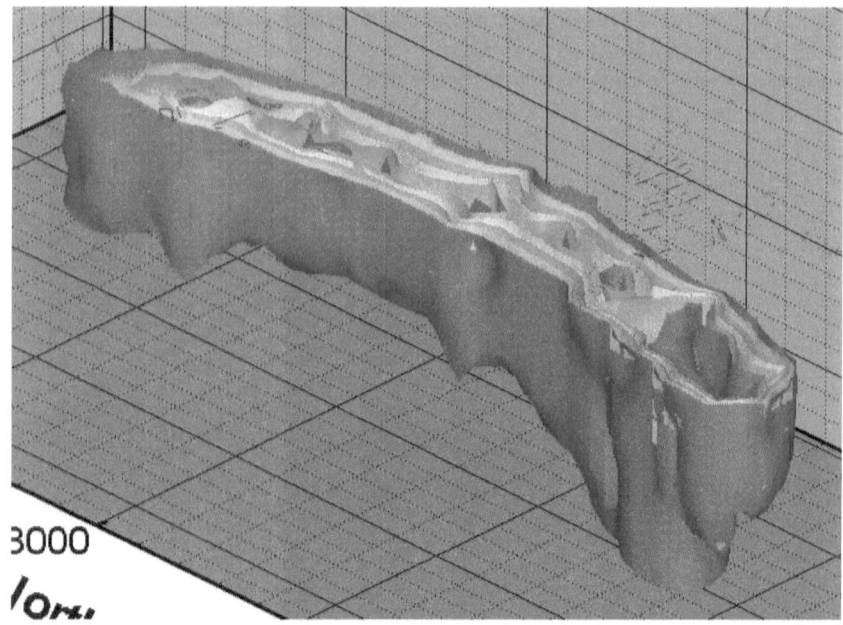

In this view, you can see how the magenta bounding polygon in the previous figure controls the overall shape of the blob. We called this plume *larva*. Data for the next plume is shown in the figure below:

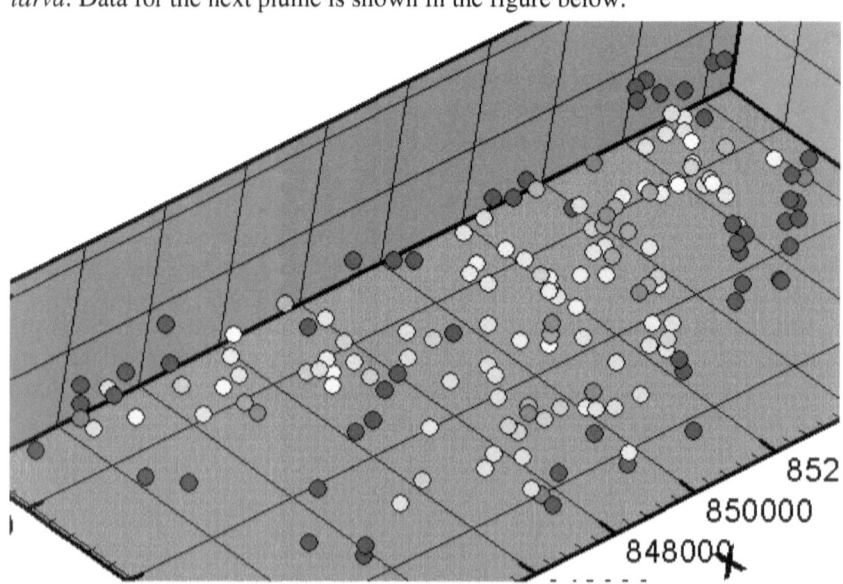

Concentration iso-surfaces for what we named the *sandal* plume shown on the cover are shown in this next figure.

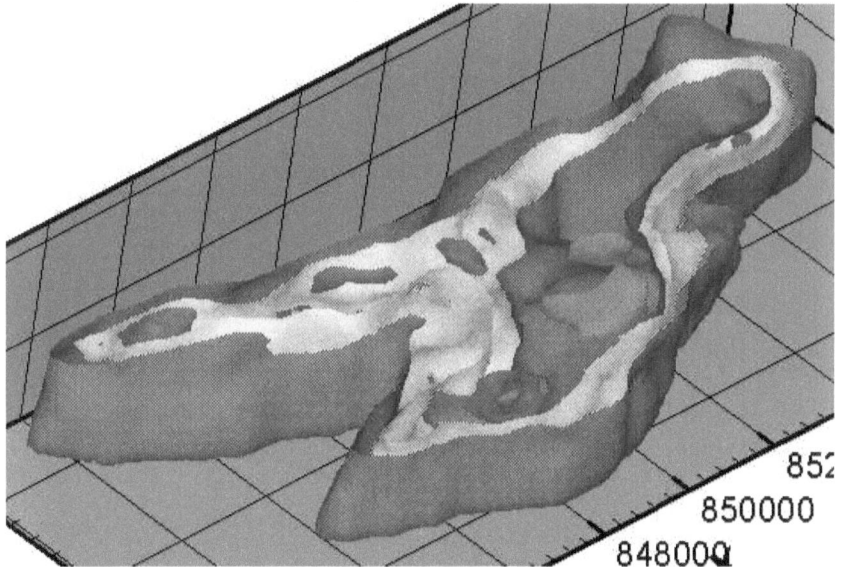

Contours for the same 3D field are:

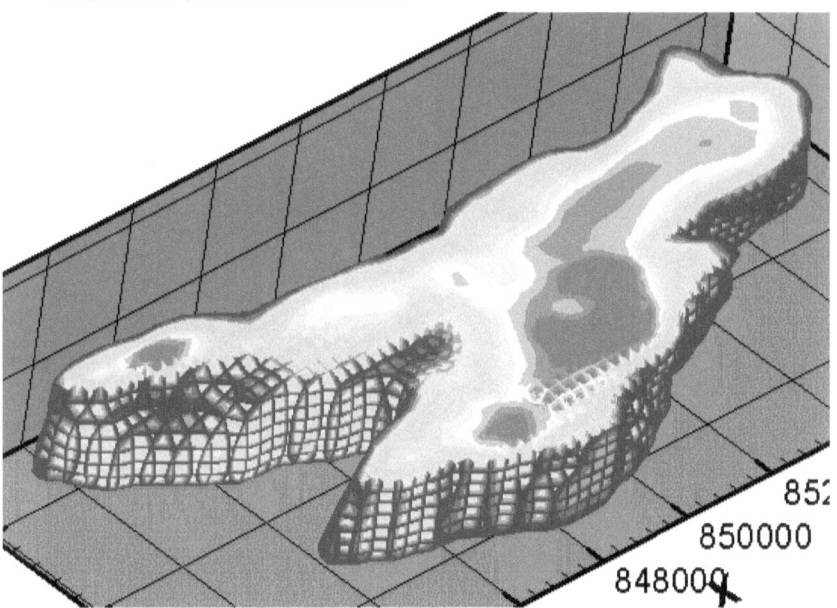

Data for the *drumstick* plume shown on page ii can be seen in this next figure:

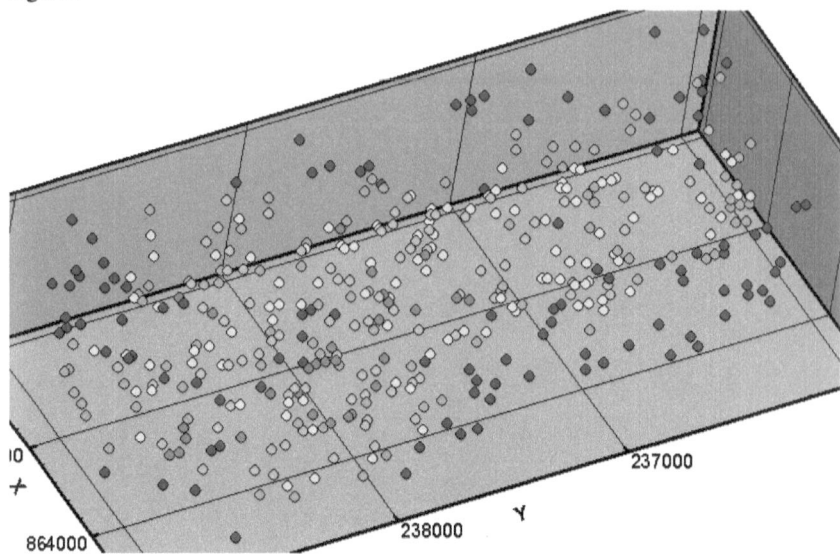

One slice of contours is shown below:

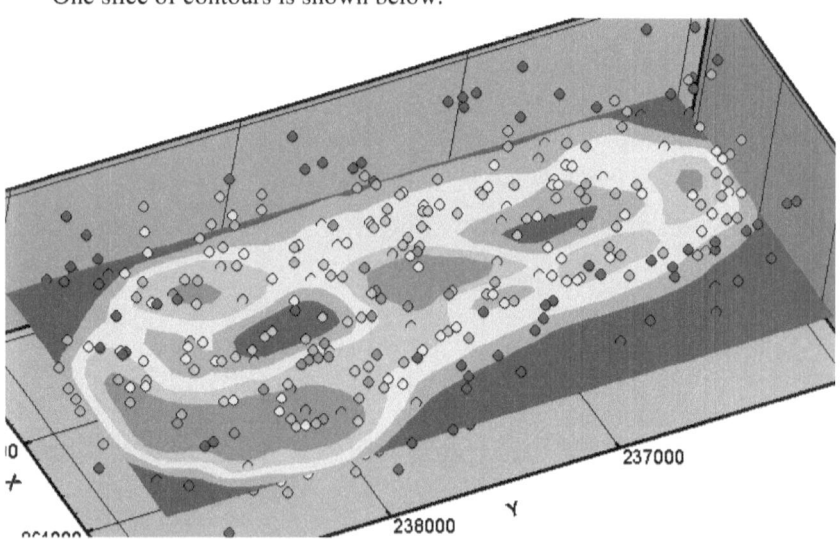

Contours in a horizontal slice through the middle of the plume and boundary for the *banana* plume are shown in this next figure along with surface details, which are quite helpful in formulating a remediation strategy:

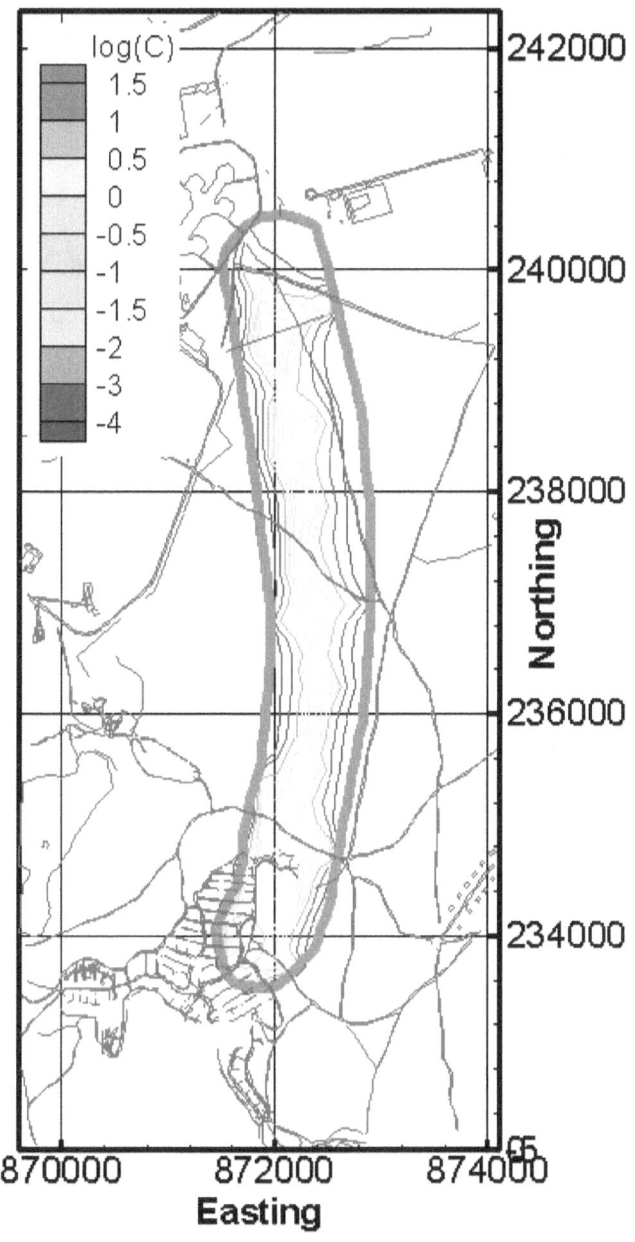

Data for the *banana* plume are shown in the figure below:

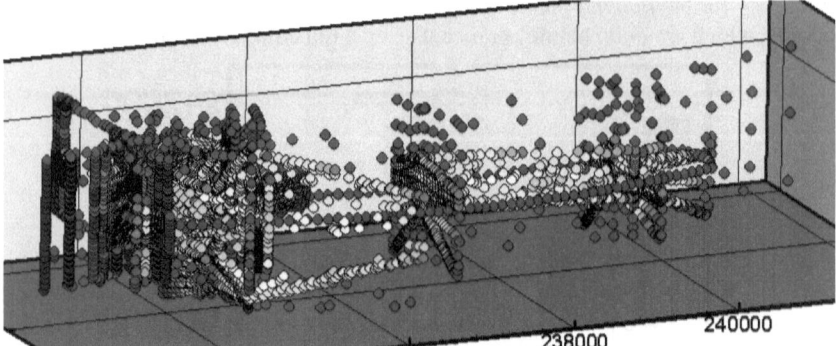

The data were concentrated near what was initially thought to be the highest concentration center of the plume.

Chapter 8. Particle Tracking of Plumes

The process of particle tracking itself has already been covered in my book entitled, *Particle Tracking*, so we won't cover that information in much detail here. The first step in tracking contaminant plumes is developing the plume, which we have covered in the preceding chapters. For our purposes here, this must be in the form of a volume, that is, a TB3 file (i.e., 3D table). The example codes (invdist.c, relax.c, and kriging.c) all produce such files, which are similar to Tecplot™ files (also produced by these same codes), only much more compact. The second step in tracking contaminant plumes is generating seeds (initial particles: position and mass), which accurately represent the plume defined by the specified volume. Add a flow model and the particle tracker to complete the process.

The first example we will consider here is the crab *claw* plume. Data plus one horizontal slice through the contours are shown in this first figure:

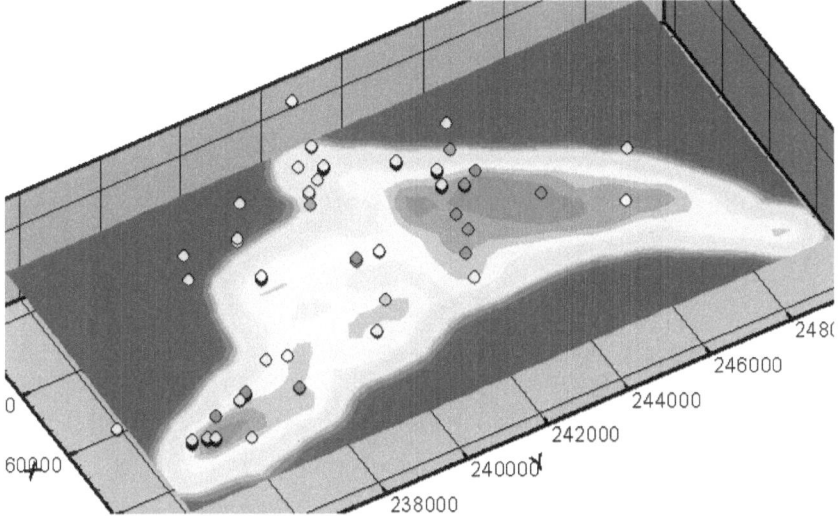

In order to create the seeds we need the volume plus a boundary polygon plus an algorithm. There are at least two ways of creating seeds to represent the concentration field: 1) uniformly distributed seeds of varying mass proportional to local concentration and 2) seeds of uniform mass distributed proportional to local concentration. Some combination of the two would be a third possibility. Generating random numbers in one dimension that have a specified probability distribution other than the normal (i.e., Gaussian) can be done, but in three dimensions and with multiple highs and lows? That is no trivial algorithm. The code (seed3d.c) can be found in the online archive in folder examples\seed3d.

Iso-surfaces of constant log(concentration) for this plume are shown in the figure below:

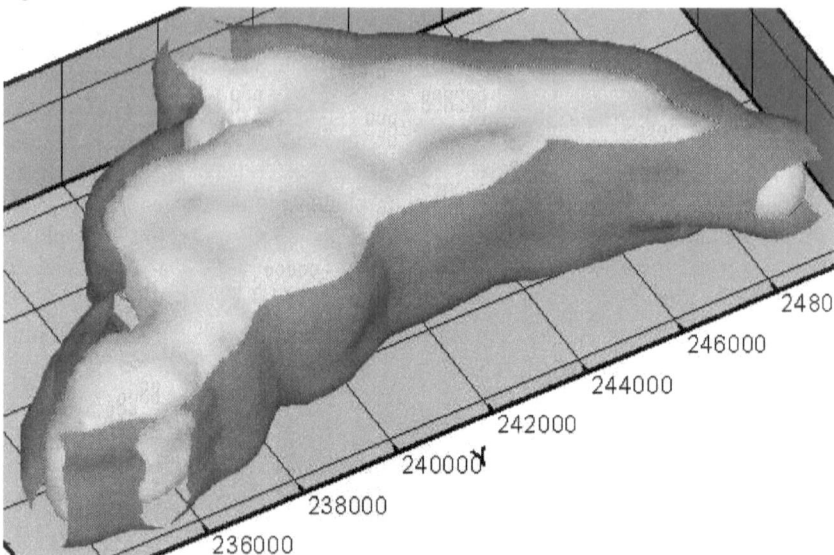

Before we describe how to create the seeds, consider the following figure showing 1775 seeds of equal mass (our first attempt):

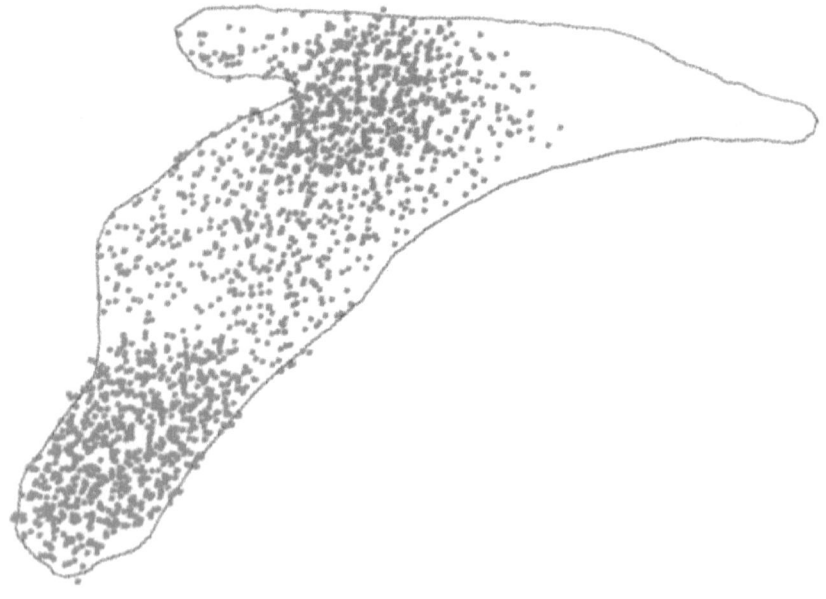

This next attempt with 50,004 seeds is much more representative of the actual plume as described by the concentration volume:

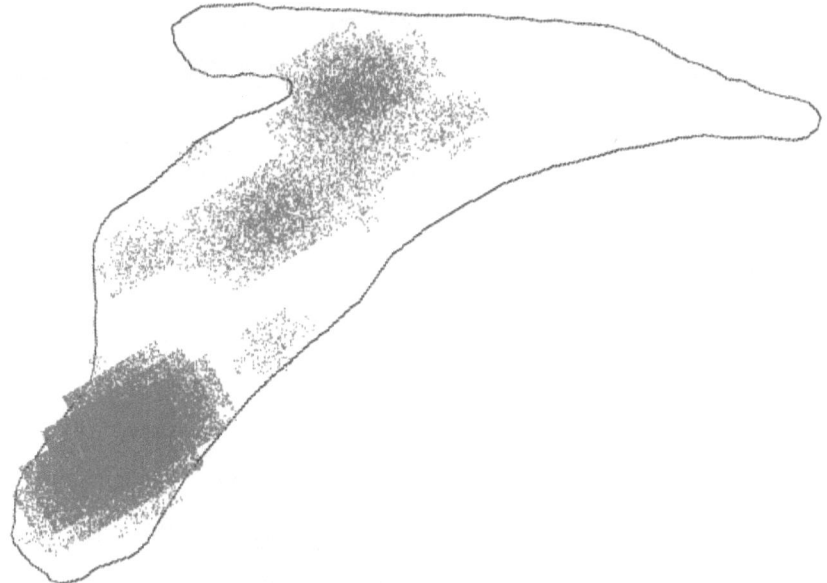

This *sprinkling* of particles is accomplished by first splitting the hexahedra (six-sided rectangular brick elements) of the volume up into tetrahedra, then filling these randomly based on the local concentration. The number of particles per tetrahedron is proportional to the concentration. The particles are scattered about inside each tetrahedra that meets the threshold concentration; otherwise, there are none. The concentration window is adjusted until the resulting number of particles roughly matches the target quantity. A side view of the seeds in the YZ plane reveals the 3D nature of this problem:

Even after the seeds are created, there's still the matter of calculating the flow field and feeding all of this information into the particle tracker. For this plume, that looks like this:

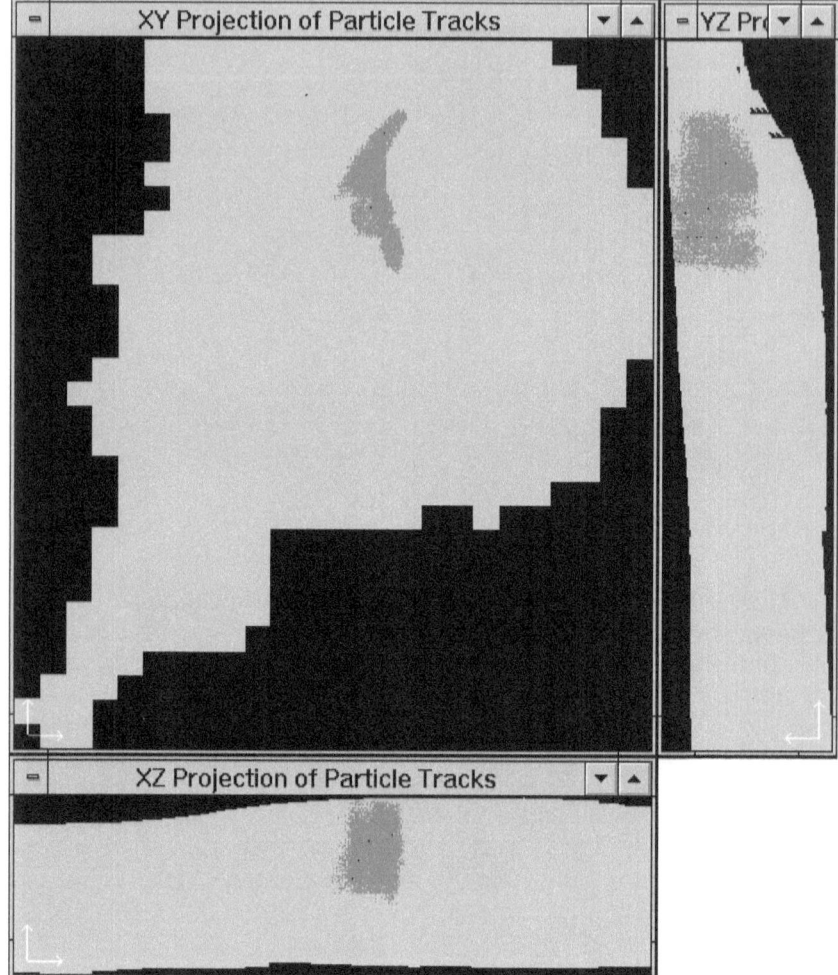

This plume was part of a larger study involving 7 plumes. This next figure shows the expected position of the plume (shown in green instead of red) after 3 years with no remediation.

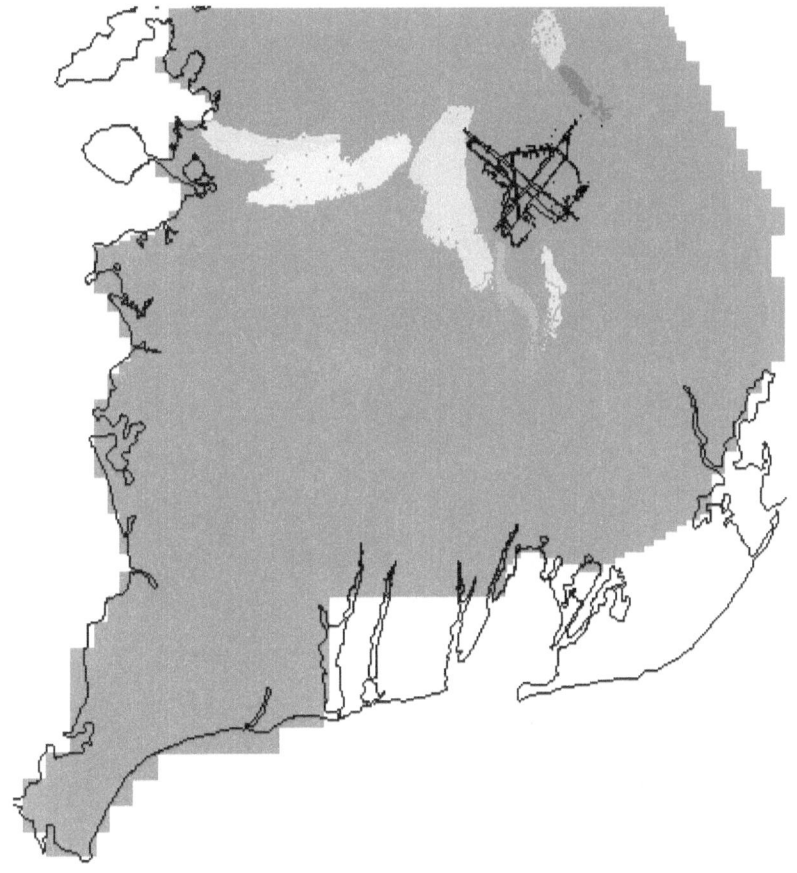

This next figure shows the expected position of the plume (in green) after 30 years without remediation.

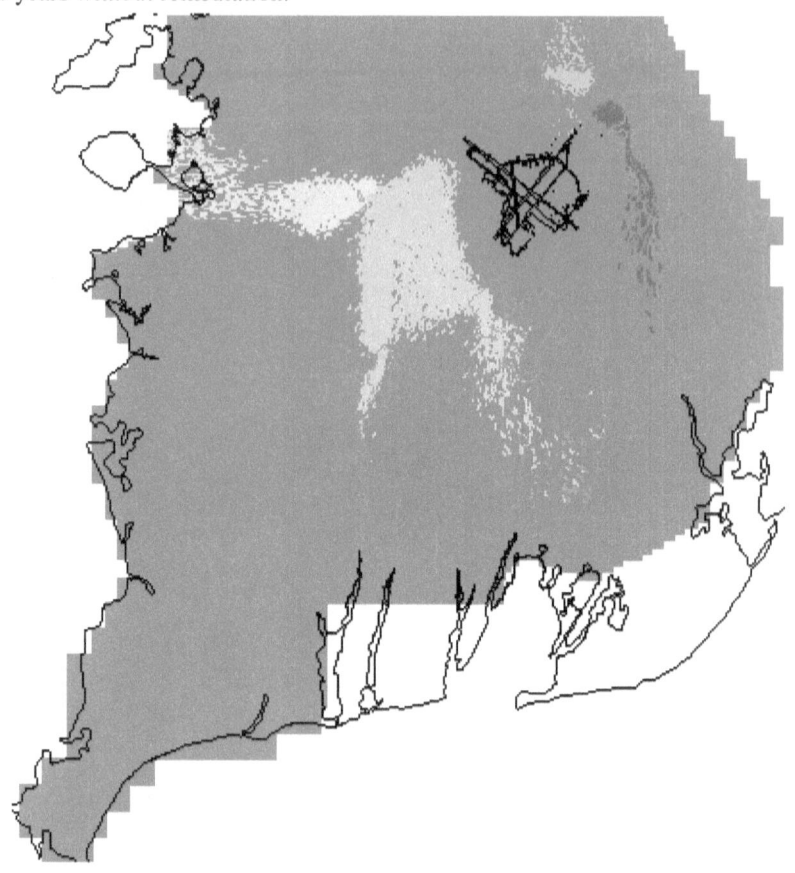

These seven plumes were remediated by installing a series of extraction wells attached to removal systems as indicated by the black +'s and o's in the following figure:

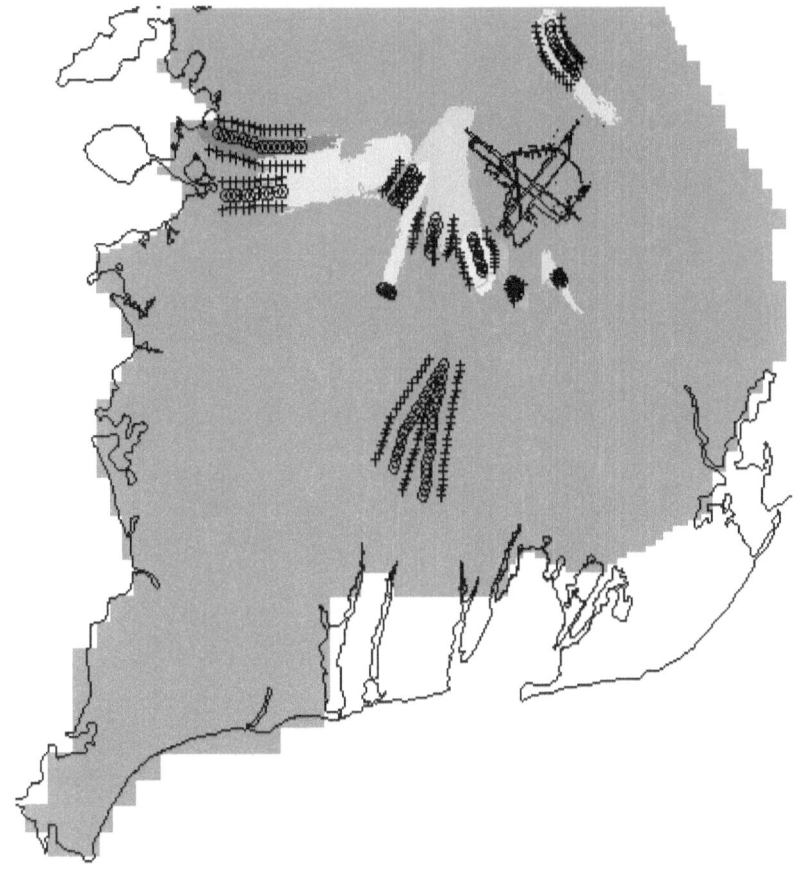

The expected results after 30 years are shown in this next figure:

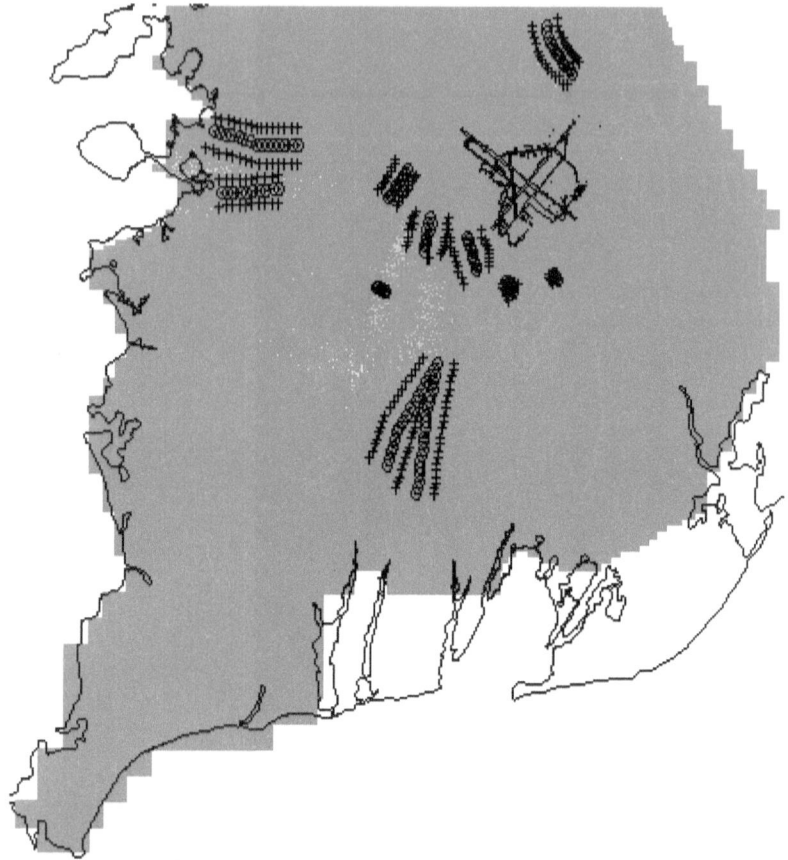

Twenty years have elapsed since this plan was implemented and the results have been quite satisfying, confirming the modeling efforts as well as the physical processes involved in capturing the contaminants.

We referred to this next plume as the *starship*:

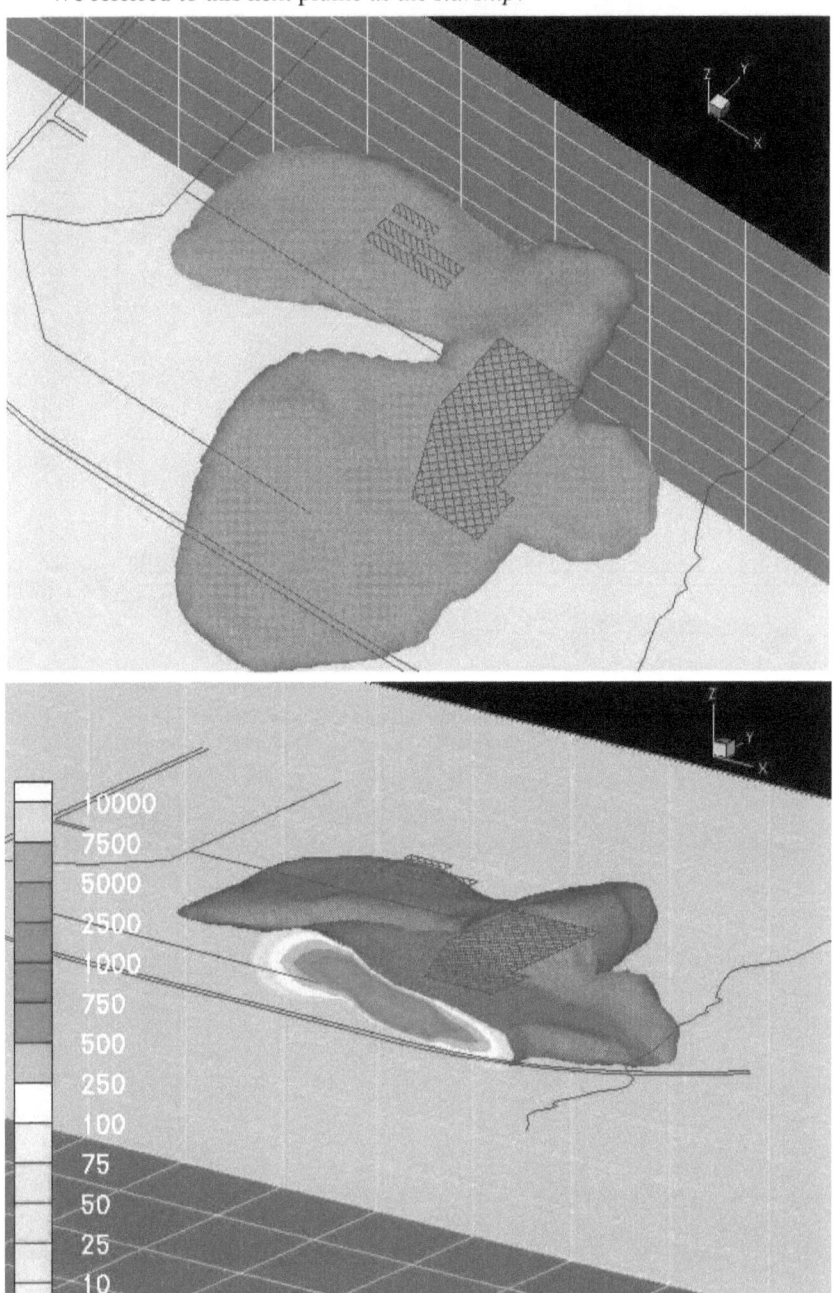

The plume was approximated by 706,042 seeds, which were sprinkled:

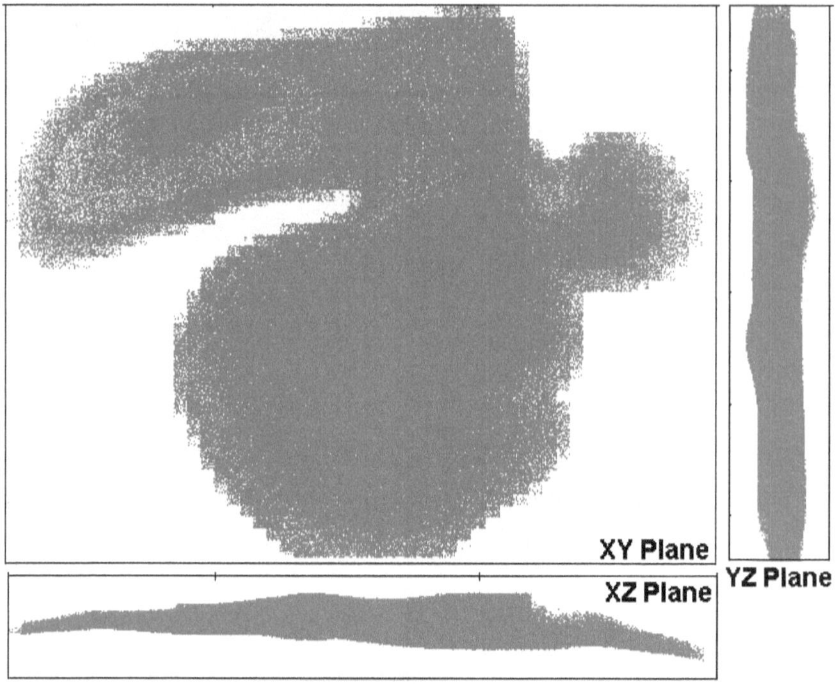

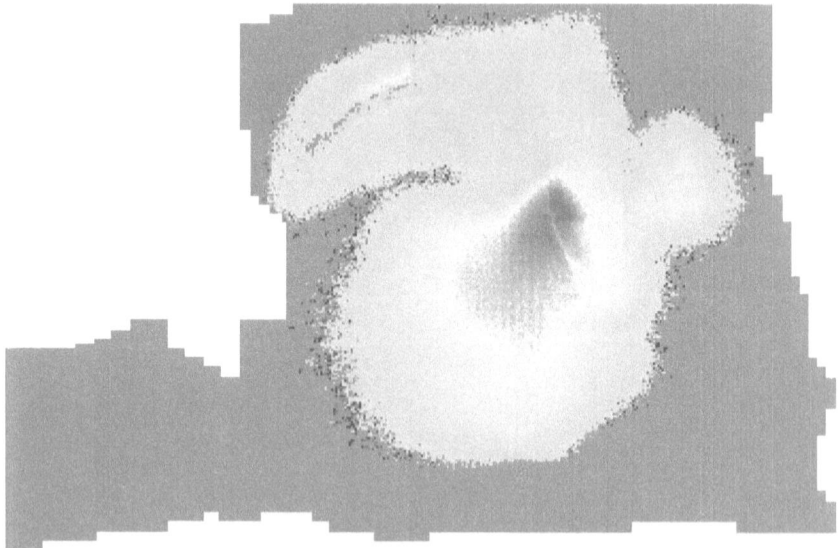

XY Plane
YZ Plane
XZ Plane

Inside the particle tracker at initialization (time=zero):

After 30 years without remediation, the expected concentrations are almost unchanged:

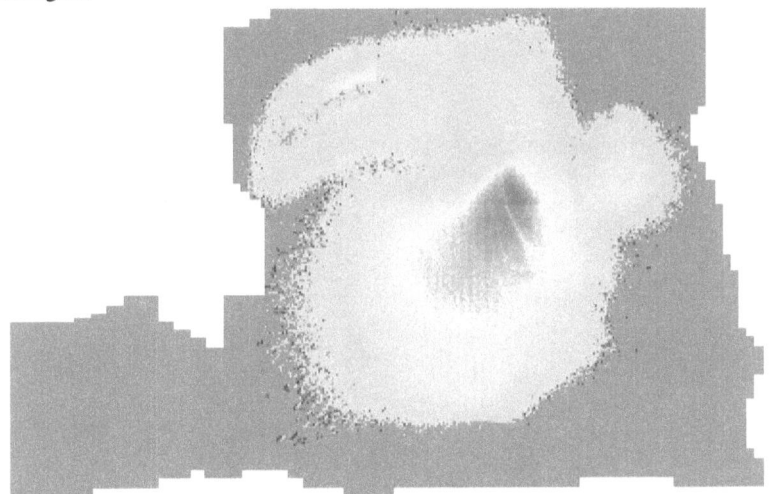

A different view of the same initial concentrations:

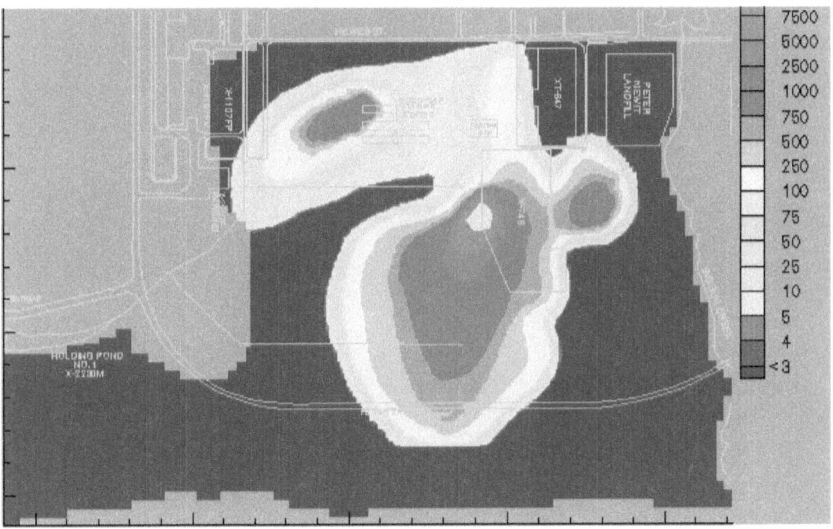

And after 30 years of remediation (first proposed approach):

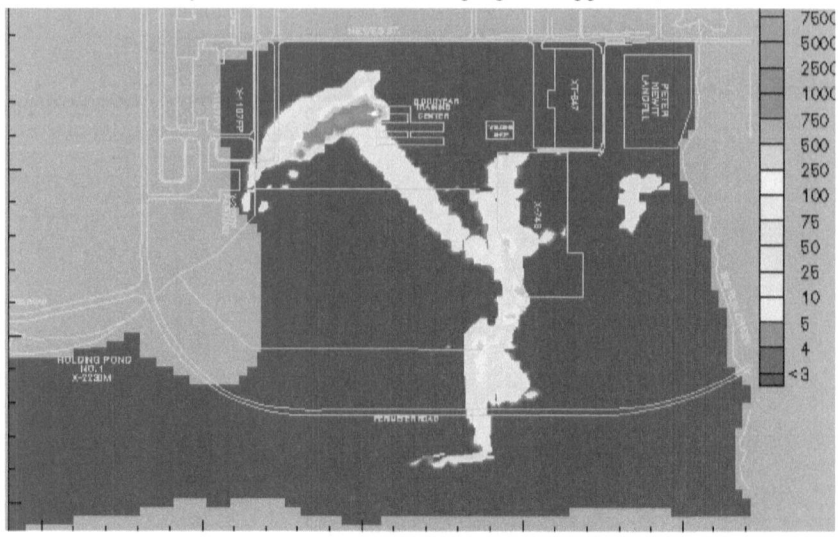

Concentration data (along with white control points) for this last plume are shown in the figure below:

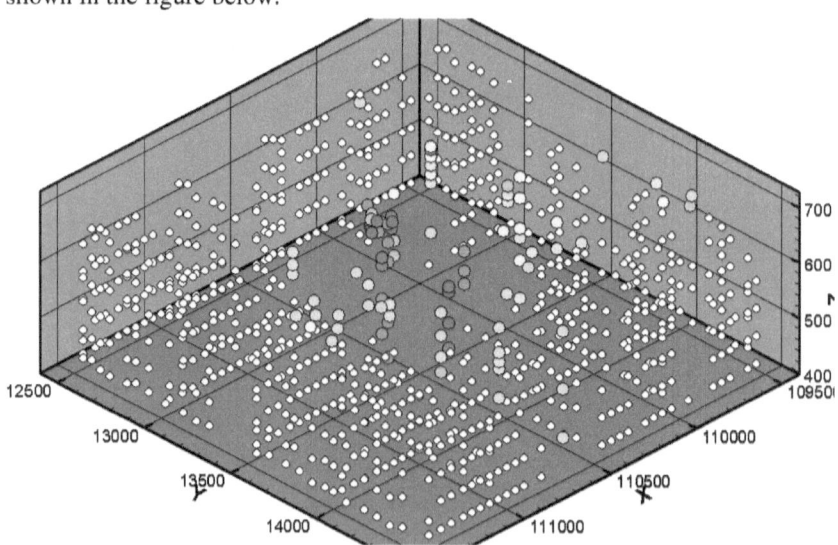

Concentration contours for this field are shown below:

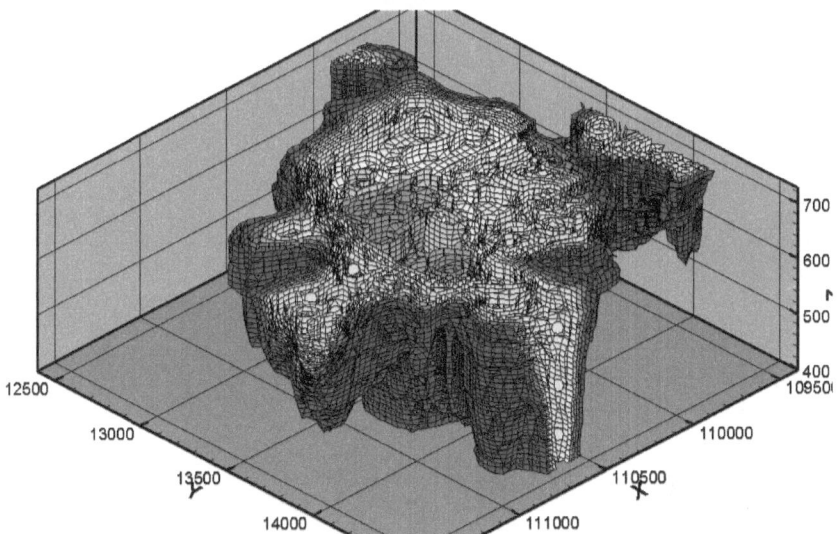

A plan view (XY plane) of the seeds:

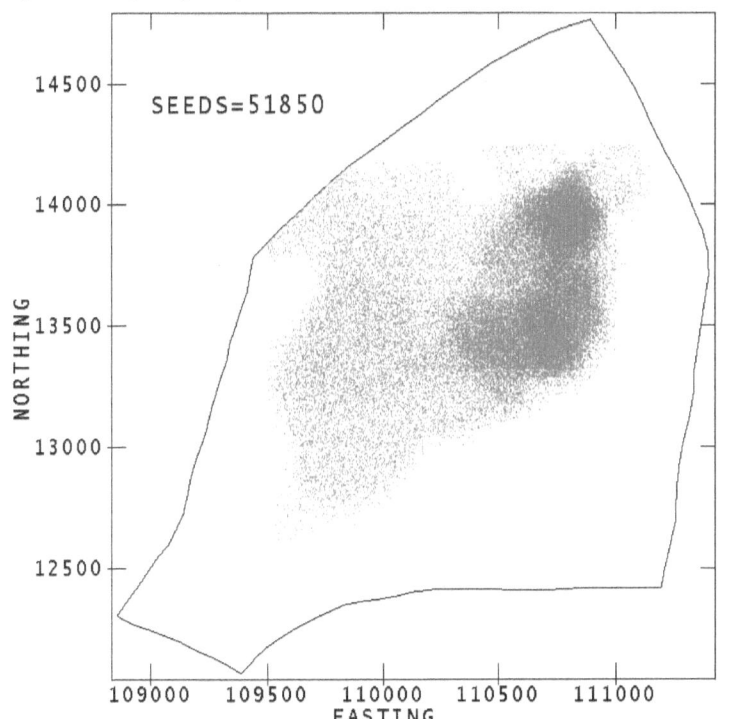

A perpendicular view of the seeds (XZ plane):

The first scenario for this plume at time t=0 is:

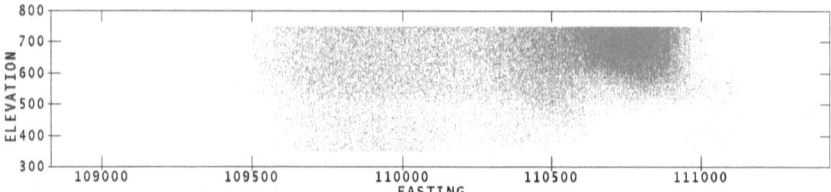

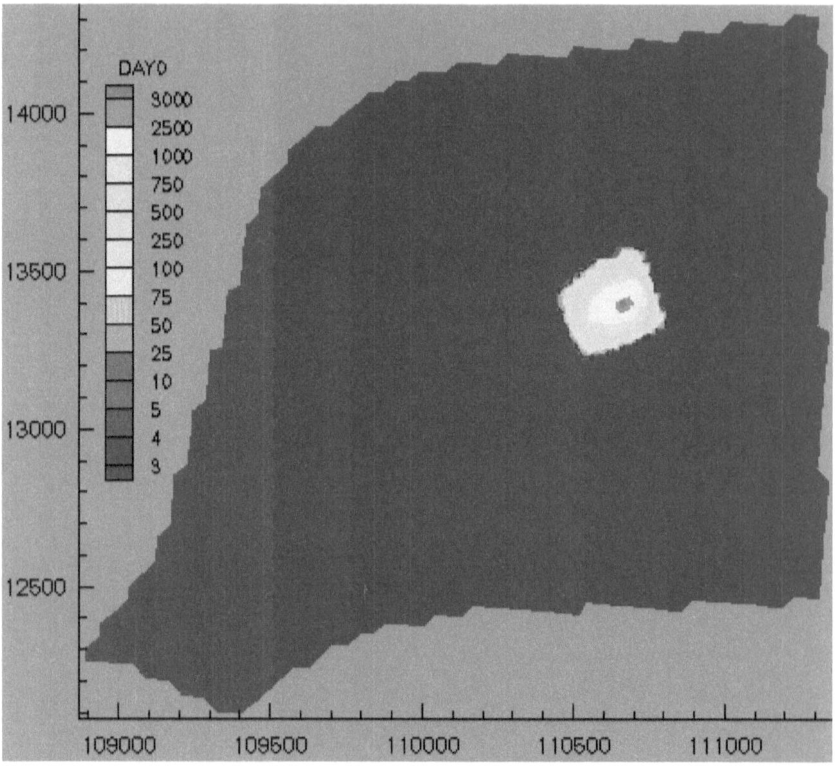

At day 6083 (16.6 years) of the original remediation plan the predicted location and concentration of the plume is:

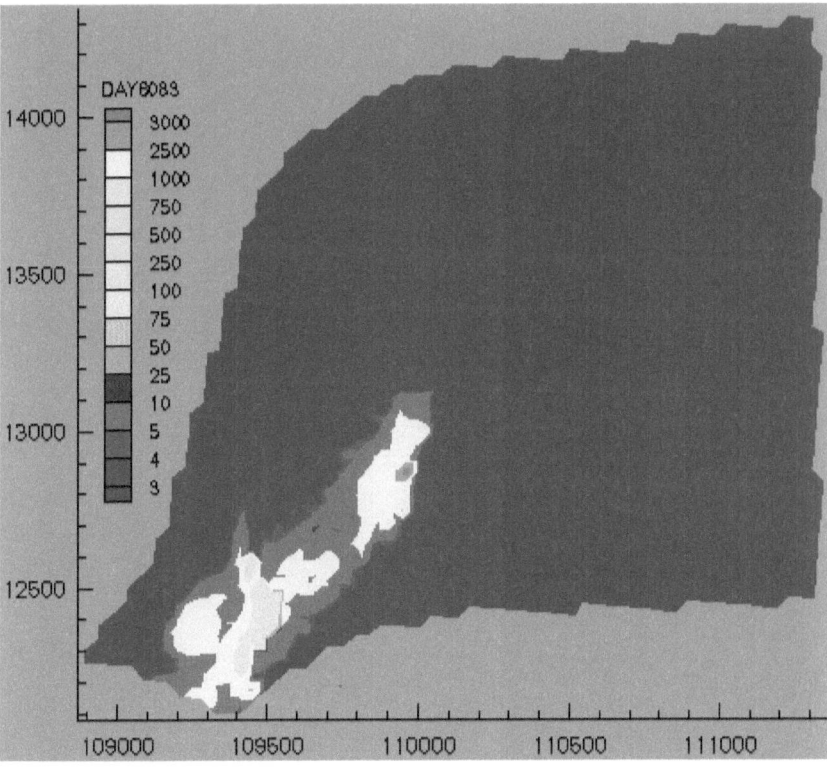

Appendix A. Inverse Distance Interpolation

The inverse distance interpolation method is one of the most useful techniques for interpolating data, especially spatially disperse and even more so irregularly spaced. In short, the closer a known point is, the more influence it should have on the local (i.e., interpolated) value. Often a power of 2.5 is applied to the distance. This can be expressed by Equation 1.1 and also the following code, which can be found in the online archive in the examples\ozone folder:

```
for(S=Z=j=0;j<Nd+Ni;j++)
  {
  D=hypot(Xd[j]-X,Yd[j]-Y);
  if(D<DBL_EPSILON)
     D=DBL_EPSILON;
  D=pow(D,2.5);
  S+=1./D;
  Z+=Zd[j]/D;
  }
Z/=S;
```

It is necessary to limit the distance so as to not divide by zero. If you're that close, it doesn't matter what the other points are. Depending on how the known points are scattered, unwanted artifacts can arise. For example, if the known points are all clumped together on one side or corner of the domain, this will produce shadows, over-emphasizing the clumped data. These unwanted artifacts can often be eliminated by considering only the closest point in each of four quadrants (or eight octants).

A mistake often made when implementing a quadrant or octant search is to use the a=arctan2(y,x) function and then a series of if(a<M_PI/4.) statements. The arctan function takes far longer to execute than the multiplications, divisions, and even raising distance to a non-integral power. Do not use arctan for this purpose. A far easier and vastly faster process is:

```
for(i=0;i<n;i++)
  {
  dX=X-Xd[i];
  dY=Y-Yd[i];
  D=hypot(dX,dY);
  q=0;
  if(dX>0.)
     q|=1;
  if(dY>0.)
     q|=2;
  if(fabs(dX)>fabs(dY))
     q|=4;
  if(D<Dq[q])
     {
     Dq[q]=D;
```

```
iQ[q]=i;
    }
}
```

The three simple (and fast) comparisons (i.e., dX>0, dY>0, and |dX|>|dY|) uniquely determine the octant (0-7) by conditionally adding 3 bits (1, 2, and 4), which can only have values 0 through 7. Using the arctan takes at least eight times as long and provides no advantage whatsoever. The code above saves the (double) distance in Dq[8] and (integer) index in iQ[8]. The summation and application of pow(D,2.5) is applied as before. Of course, depending on where you are in the field, some of the octants may be empty, which is why I first fill the array iQ with minus ones.

Inverse distance code for 2D and 3D can be found in the online archive in folder examples\invdist and also several of the other examples, including ozone. Use of octants can be selected by changing the corresponding conditional compilation statements. Typical 2D results are shown in the following figure:

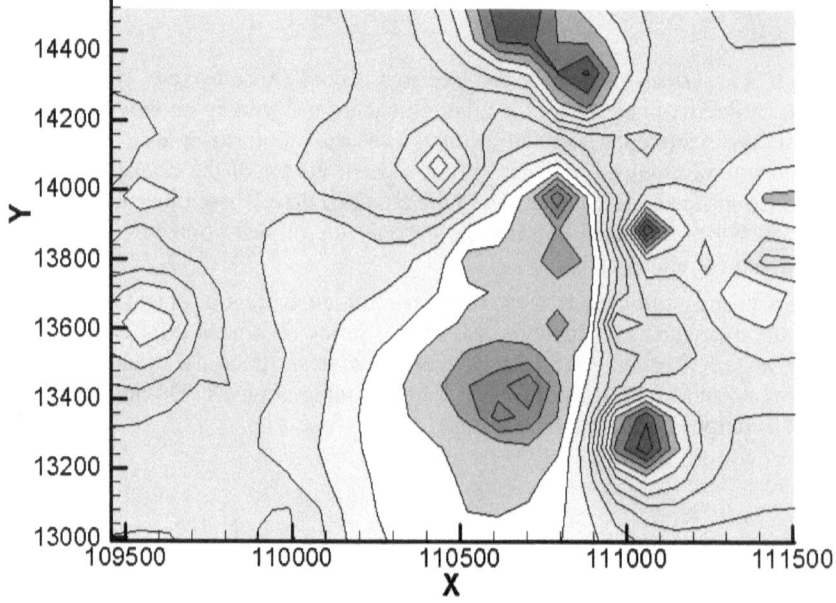

Inverse Distance in Three Dimensions

While there is sometimes disparity in transport or spreading in the two horizontal directions (X and Y), more often there is disparity between the horizontal and vertical. When this is the case, inverse distance interpolation must be modified to account for this. Most of the time this can be simply adjusted with a scaling factor, as in:

```
D=sqrt(dX*dX+dY*dY+A*dZ*dZ);
```

The code (invdist.c) will read in the data and determine if it is 2D or 3D and interpolate accordingly. The domain is optionally expanded by some percentage, typically 5% or 10%. Typical 3D data are shown in the following figure:

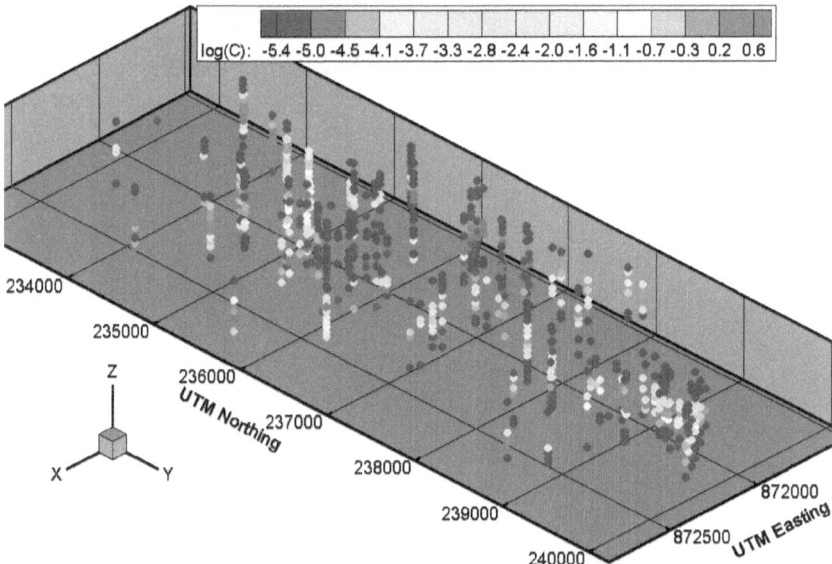

Three-dimensional data (i.e., X,Y,Z,C) is a volume and must be sliced or converted to iso-surfaces in order to visualize it. Both Tecplot™ and TP2 can perform this task. One slice of contours is shown in this next figure:

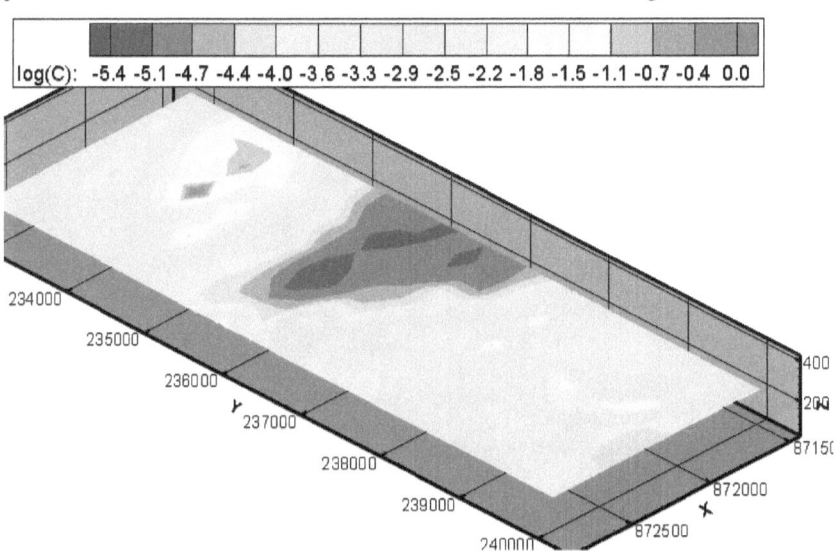

Note that both Tecplot™ and TP2 will perform the inverse distance method for you and also display the data. Several options are available. TP2 also performs the relaxation method described in Appendix B. Tecplot™ will relax a 2D or 3D field for you, but not with the control point strategy, which is often essential in obtaining acceptable results. Tecplot™ is an excellent commercial product. TP2 is available free online at the address provided in the Forward.

Appendix B. Relaxation Method

The relaxation method is sometimes called *smoothing*, for it has this effect, especially if viewed sequentially, were one to create an animation of the process. The basis for this method is actually solving Laplace's equation. In two dimension, this is expressed:

$$\frac{\partial^2 \varphi}{\partial x^2} + \frac{\partial^2 \varphi}{\partial y^2} = 0 \tag{B.1}$$

A graphical representation for the finite difference approximation of this partial differential equation is:

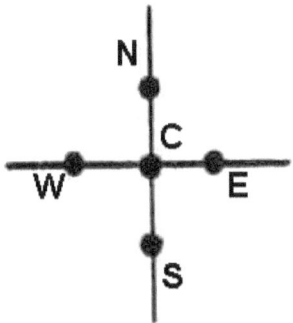

The corresponding finite difference equation for uniformly spaced points is:

$$\frac{(\varphi_N + \varphi_S + \varphi_E + \varphi_W - 4\varphi_C)}{\Delta^2} = 0 \tag{B.2}$$

This can be rearranged to form:

$$\varphi_C \frac{(\varphi_N + \varphi_S + \varphi_E + \varphi_W)}{4} \tag{B.3}$$

This is the same as averaging the four surrounding points. The same thing extends to three dimensions:

$$\varphi_C \frac{(\varphi_N + \varphi_S + \varphi_E + \varphi_W + \varphi_{UP} + \varphi_{DOWN})}{6} \tag{B.4}$$

This process is very simple to implement. You typically have control points or locations where you know the value, for instance, field data or boundaries. This is handled by a Boolean index for each point (0 or 1). The relaxation method requires a lot more memory than the inverse distance method described in Appendix A. Depending on the number of data points, one method may take considerably more computational time than the other. The relaxation method performs much better with more calculation points (i.e., finer granularity) than the inverse distance method, which result is independent of the number of calculation points. The code (relax.c) can be found in the online archive in folder examples\relax. The same data files from folder examples\invdist can be used. The results for the 2D example are:

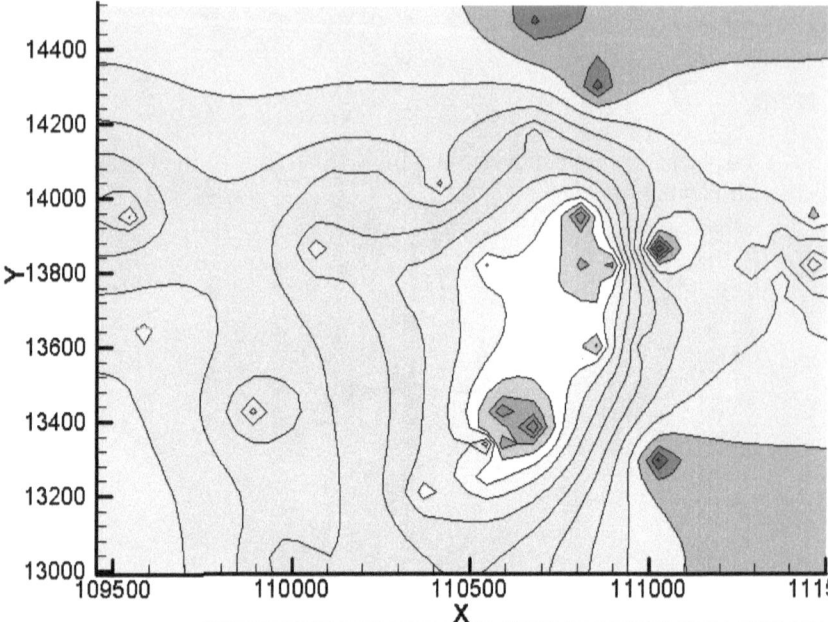

You can easily add scale disparity to the calculation, as this is simply a weighting (more or less) of the points in one direction (East/West, North/South, or Up/Down). The relaxation loop is quite simple:

```
for(w=z=0;z<Nz;z++)
  {
  for(y=0;y<Ny;y++)
    {
    for(x=0;x<Nx;x++,w++)
      {
      if(fix[w])
         continue;
      Cr[w]=n=0;
      if(x>0)
         {
         Cr[w]+=Cr[w-1];
         n++;
         }
      if(x<Nx-1)
         {
         Cr[w]+=Cr[w+1];
         n++;
```

```
            }
         if(y>0)
            {
            Cr[w]+=Cr[w-Nx];
            n++;
            }
         if(y<Ny-1)
            {
            Cr[w]+=Cr[w+Nx];
            n++;
            }
         Cr[w]/=n;
         }
      }
   }
```
The 3D example results are shown in the figure below:

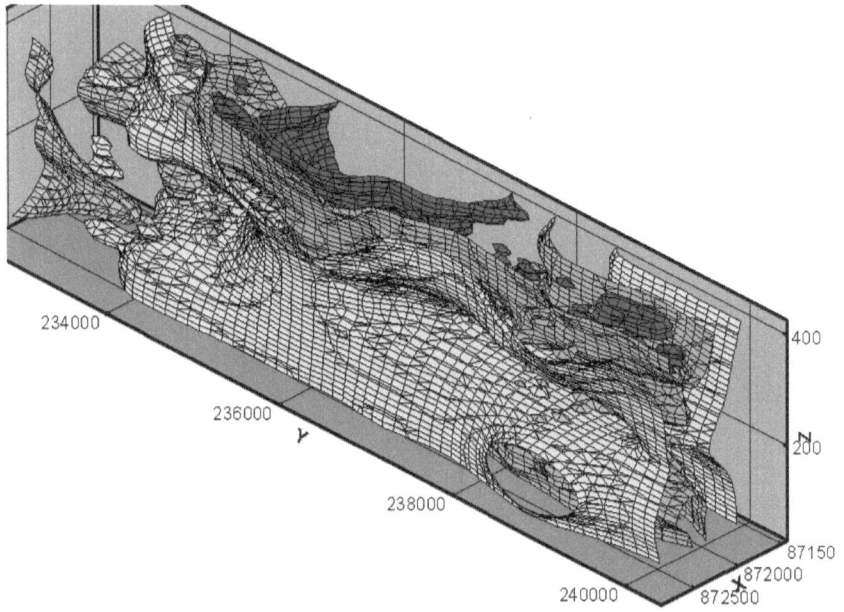

Appendix C. Kriging

Kriging is a form of Gaussian regression or interpolation used primarily in geostatistics. The original method was developed by Matheron[40] and based on the work of Krige.[41] There are actually a variety of formulas and techniques associated with the term *kriging*, making this designation rather ambiguous. Even a cursory search of the Internet will produce a dozen formulas, many of which (e.g., cos) bear no resemblance to the concept of Gaussian. Illustrations in one dimension show each data point blending into the next, but this is far removed from two- and three-dimensional hydrogeological data such as we have been discussing in this text.

Tecplot™ contains an implementation of this method that works well for some data but not for others. The same could be said for every other application I have used, so this is by no means a criticism of Tecplot™. There is a comparison of inverse distance and kriging in Chapter 6. In that particular case, kriging is a complete flop. Still, we won't discard the concept altogether.

Recall from Appendix B that the relaxation method is like solving Laplace's equation, which is steady-state conduction or diffusion. The control points (i.e., locations where the value is known) might be viewed as point sources (locations of fixed temperature, potential, or concentration). The 2D/3D field arises from the point sources. In fact, this is the same solution we would get if we were to solve a heat conduction or diffusion problem with the data points as fixed values. There is no particular function associated with the intervening spaces, except that it must satisfy Laplace's equation.

If we want kriging to be Gaussian and we consider these two approaches (kriging and relaxation) together, this leads to the question: Why not solve the problem of point sources having exponential decay? This gets us back to some of the equations you will find online, in particular $\Sigma C_I / \exp(-r^2/\lambda^2)$. We must first select a length scale, λ, and then determine the contribution of each data point (C_I). We could build a set of NxN simultaneous linear equations and simply solve for the values. That might work for a few dozen data points, perhaps even 100. Given 1000 data points, the chance of obtaining meaningful results from direct solution of 1000x1000 linear equations is nil.

Noting that the current point will always have more influence than any other (i.e., $\exp(0)=1$, $\exp(<0)<1$), the diagonal elements of the resulting matrix will always be the largest terms, if not strictly dominating. Also, we note that

---

[40] Georges François Paul Marie Matheron (1930–2000) was French mathematician and civil engineer of mines, known as the founder of geostatistics and a co-founder (together with Jean Serra) of mathematical morphology.

[41] Danie Gerhardus Krige (1919–2013) South African statistician and mining engineer who pioneered the field of geostatistics and was professor at the University of the Witwatersrand.

concentrations and conductivities might be zero, but are not negative, which forces each contribution to be positive. These combine to make iterative solution of the resulting system of simultaneous equations tractable. Because the diagonal elements are not strictly dominant, iterations tend to over-shoot. Rather than using Successive Over Relaxation (SOR), we use under-relaxation (not related to Appendix B). The iterations also tend to oscillate; so we force them to converge with progressive dampening. The inner loop is quite small:

```
for(w=0;w<64;w++)
  {
  for(d=0;d<Nd;d++)
    {
    C=Cd[d];
    for(c=0;c<Nd;c++)
      if(c-d)
        C-=Ck[d]*Bd[Nd*d+c];
    Ck[d]=fmax(Ck[d]/2.,fmin(2.*Cd[d],
(w*Ck[d]+C)/(w+1)));
    }
  }
```

Progressive dampening is implemented by the term w/(w+1). The number of iterations may require adjustment, here 64 are taken. The influence factors (Bd) need be calculated only once, as these are the exp($-r^2/\lambda^2$) terms, which don't change with iteration. The code (kriging.c) is in the examples\kriging folder.

This modified kriging process yields very smooth results—too smooth in some cases. The 2D data set from the examples\invdist is shown below:

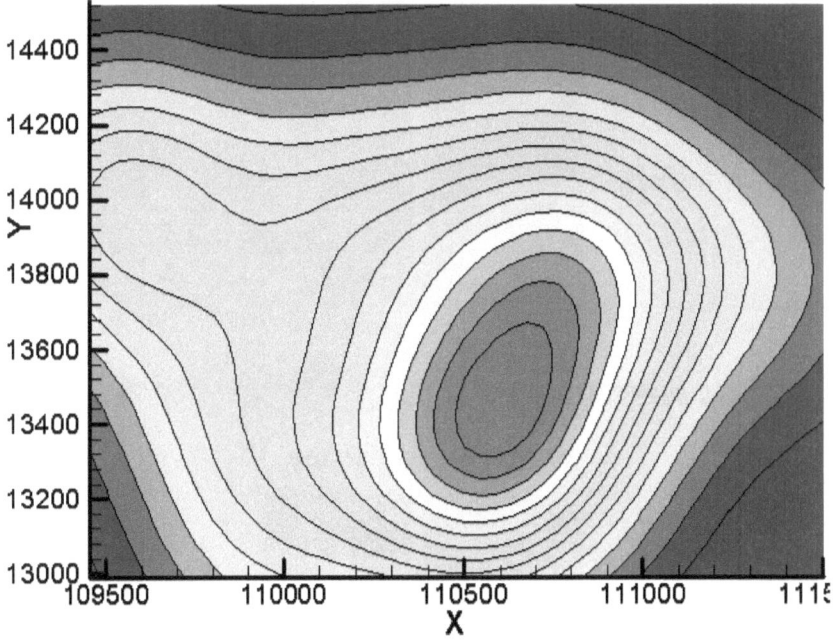

The 3D data set from the examples\invdist is shown in this next figure:

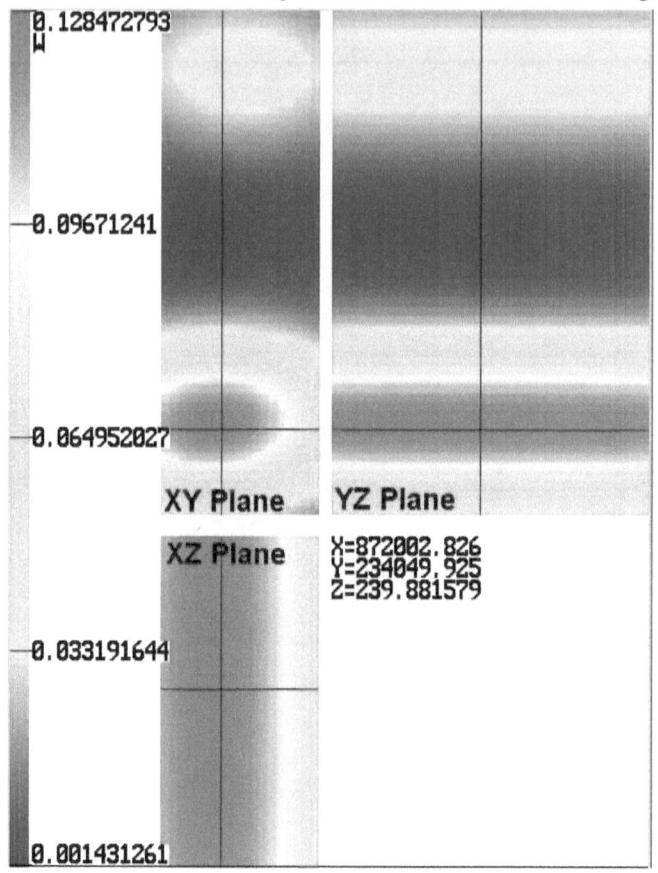

Appendix D. BMP to GIF Conversion

The code required to create a single frame and save it as a BMP file is rather small (see for instance, ozone.c in examples\ozone). Creating a GIF is considerably more complex (but much less complex than a JPEG). I have provided a utility with source code (bmp2gif) that uses wild card file matching (e.g., ozone*.bmp) to combine multiple images into a single animated GIF. You can download it any time from my web site. It is included in the examples for *Compression & Encryption*:

> https://dudleybenton.altervista.org/software/Compression.zip

and also for *Orthogonal Functions*:

> https://dudleybenton.altervista.org/software/Orthogonal Functions/orthogonal_function_examples.zip

also by D. James Benton

3D Articulation: Using OpenGL, ISBN-9798596362480, Amazon, 2021 (book 3 in the 3D series).

3D Models in Motion Using OpenGL, ISBN-9798652987701, Amazon, 2020 (book 2 in the 3D series.

3D Rendering in Windows: How to display three-dimensional objects in Windows with and without OpenGL, ISBN-9781520339610, Amazon, 2016 (book 1 in the 3D series).

A Synergy of Short Stories: The whole may be greater than the sum of the parts, ISBN-9781520340319, Amazon, 2016.

Azeotropes: Behavior and Application, ISBN-9798609748997, Amazon, 2020.

bat-Elohim: Book 3 in the Little Star Trilogy, ISBN-9781686148682, Amazon, 2019.

Boilers: Performance and Testing, ISBN: 9798789062517, Amazon 2021.

Combined 3D Rendering Series: 3D Rendering in Windows®, 3D Models in Motion, and 3D Articulation, ISBN-9798484417032, Amazon, 2021.

Complex Variables: Practical Applications, ISBN-9781794250437, Amazon, 2019.

Compression & Encryption: Algorithms & Software, ISBN-9781081008826, Amazon, 2019.

Computational Fluid Dynamics: an Overview of Methods, ISBN-9781672393775, Amazon, 2019.

Computer Simulation of Power Systems: Programming Strategies and Practical Examples, ISBN-9781696218184, Amazon, 2019.

Contaminant Transport: A Numerical Approach, ISBN-9798461733216, Amazon, 2021.

CPUnleashed! Tapping Processor Speed, ISBN-9798421420361, Amazon, 2022.

Curve-Fitting: The Science and Art of Approximation, ISBN-9781520339542, Amazon, 2016.

Death by Tie: It was the best of ties. It was the worst of ties. It's what got him killed., ISBN-9798398745931, Amazon, 2023.

Differential Equations: Numerical Methods for Solving, ISBN-9781983004162, Amazon, 2018.

Equations of State: A Graphical Comparison, ISBN-9798843139520, Amazon, 2022.

Evaporative Cooling: The Science of Beating the Heat, ISBN-9781520913346, Amazon, 2017.

Forecasting: Extrapolation and Projection, ISBN-9798394019494, Amazon 2023.

Heat Engines: Thermodynamics, Cycles, & Performance Curves, ISBN-9798486886836, Amazon, 2021.

Heat Exchangers: Performance Prediction & Evaluation, ISBN-9781973589327, Amazon, 2017.

Heat Recovery Steam Generators: Thermal Design and Testing, ISBN-9781691029365, Amazon, 2019.

Heat Transfer: Heat Exchangers, Heat Recovery Steam Generators, & Cooling Towers, ISBN-9798487417831, Amazon, 2021.

Heat Transfer Examples: Practical Problems Solved, ISBN-9798390610763, Amazon, 2023.

The Kick-Start Murders: Visualize revenge, ISBN-9798759083375, Amazon, 2021.

Jamie2: Innocence is easily lost and cannot be restored, ISBN-9781520339375, Amazon, 2016-18.

Kyle Cooper Mysteries: Kick Start, Monte Carlo, and Waterfront Murders, ISBN-9798829365943, Amazon, 2022.

The Last Seraph: Sequel to Little Star, ISBN-9781726802253, Amazon, 2018.

Little Star: God doesn't do things the way we expect Him to. He's better than that! ISBN-9781520338903, Amazon, 2015-17.

Living Math: Seeing mathematics in every day life (and appreciating it more too), ISBN-9781520336992, Amazon, 2016.

Lost Cause: If only history could be changed..., ISBN-9781521173770, Amazon, 2017.

Mass Transfer: Diffusion & Convection, ISBN-9798702403106, Amazon, 2021.

Mill Town Destiny: The Hand of Providence brought them together to rescue the mill, the town, and each other, ISBN-9781520864679, Amazon, 2017.

Monte Carlo Murders: Who Killed Who and Why, ISBN-9798829341848, Amazon, 2022.

Monte Carlo Simulation: The Art of Random Process Characterization, ISBN-9781980577874, Amazon, 2018.

Nonlinear Equations: Numerical Methods for Solving, ISBN-9781717767318, Amazon, 2018.

Numerical Calculus: Differentiation and Integration, ISBN-9781980680901, Amazon, 2018.

Numerical Methods: Nonlinear Equations, Numerical Calculus, & Differential Equations, ISBN-9798486246845, Amazon, 2021.

Orthogonal Functions: The Many Uses of, ISBN-9781719876162, Amazon, 2018.

Overwhelming Evidence: A Pilgrimage, ISBN-9798515642211, Amazon, 2021.

Particle Tracking: Computational Strategies and Diverse Examples, ISBN-9781692512651, Amazon, 2019.

Power Plant Performance Curves: for Testing and Dispatch, ISBN-9798640192698, Amazon, 2020.

Practical Linear Algebra: Principles & Software, ISBN-9798860910584, Amazon, 2023.

Props, Fans, & Pumps: Design & Performance, ISBN-9798645391195, Amazon, 2020.

Remediation: Contaminant Transport, Particle Tracking, & Plumes, ISBN-9798485651190, Amazon, 2021.
ROFL: Rolling on the Floor Laughing, ISBN-9781973300007, Amazon, 2017.
Seminole Rain: You don't choose destiny. It chooses you, ISBN-9798668502196, Amazon, 2020.
Septillionth: 1 in 10^{24}, ISBN-9798410762472, Amazon, 2022.
Software Development: Targeted Applications, ISBN-9798850653989, Amazon, 2023.
Software Recipes: Proven Tools, ISBN-9798815229556, Amazon, 2022.
Steam 2020: to 150 GPa and 6000 K, ISBN-9798634643830, Amazon, 2020.
Thermochemical Reactions: Numerical Solutions, ISBN-9781073417872, Amazon, 2019.
Thermodynamic and Transport Properties of Fluids, ISBN-9781092120845, Amazon, 2019.
Thermodynamic Cycles: Effective Modeling Strategies for Software Development, ISBN-9781070934372, Amazon, 2019.
Thermodynamics - Theory & Practice: The science of energy and power, ISBN-9781520339795, Amazon, 2016.
Version-Independent Programming: Code Development Guidelines for the Windows® Operating System, ISBN-9781520339146, Amazon, 2016.
The Waterfront Murders: As you sow, so shall you reap, ISBN-9798611314500, Amazon, 2020.
Weather Data: Where To Get It and How To Process It, ISBN-9798868037894, Amazon, 2023.

www.ingramcontent.com/pod-product-compliance
Lightning Source LLC
Chambersburg PA
CBHW030715220526
45463CB00005B/2057